Energy and Economic Development in India

R. K. Pachauri

The Praeger Special Studies program—utilizing the most modern and efficient book production techniques and a selective worldwide distribution network—makes available to the academic, government, and business communities significant, timely research in U.S. and international economic, social, and political development.

Energy and Economic Development in India

Praeger Special Studies in International Economics and Development

Praeger Publishers   New York   London   

Library of Congress Cataloging in Publication Data

Pachauri, R   K
   Energy and economic development in India.

   (Praeger special studies in international economics and development)
   Bibliography: p.
   Includes index.
   1. Power resources—India. 2. India—Economic conditions—1947-   I. Title.
HD9502.I42P32   1977   333.7   77-12718
ISBN 0-03-022371-7

Robert Manning Strozier Library

FEB 22 1978

Tallahassee, Florida

PRAEGER SPECIAL STUDIES
200 Park Avenue, New York, N.Y., 10017, U.S.A.

Published in the United States of America in 1977
by Praeger Publishers,
A Division of Holt, Rinehart and Winston, CBS, Inc.

789   038   987654321

© 1977 by Praeger Publishers

All rights reserved

Printed in the United States of America

In ever fond memory of
my brother Lt. Col. V. K. Pachauri—
scholar, soldier, and sportsman
and above all the most loving and
warmest of human beings

PREFACE

The title of this book suggests that what will be served will be a goulash cooked with a variety of commonly used ingredients. In this respect, perhaps, the title is appropriate, for like a politician's set of promises before an election, there is something in it for everyone—but perhaps there the analogy ends. For what is served in these pages is perhaps insipid and unspiced, and even though I started with a very precise recipe, it was modified as regularly as a cook varies the recipe while stirring the pot. Hence, let me state right here that my image of this book and what I should put in it kept changing as the pages progressed.

The problems of energy in developing countries—and particularly those that are major importers of oil, such as India—are acute not so much because of a poor resource base in various forms of energy but more because of the major neglect suffered by the energy sector in developing economies. The author's interest and research in the field of energy efforts date back over six years (the results of which are in a book published by Praeger). In 1976, the author organized a National Seminar on Energy at the Administrative Staff College of India at Hyderabad. This seminar was addressed by the cabinet ministers for Energy and Petroleum from the Indian government, and participants included leading technologists, scientists, economists, sociologists, government officials, and others connected with the field of energy in India. It soon became apparent (both during and as a result of the deliberations of the seminar) that the problems of energy cut across interdisciplinary boundaries and had to be viewed in the wider perspective of economic development and change in a tradition-bound society. It was also felt by all those with whom the author exchanged views that even though significant work had been done by a number of individuals and committees in analyzing the complex area of energy in the country, hardly anything had been written to present (in one volume) a comprehensive study of energy problems within a broad framework. There is a paucity of various kinds of data essential for an in-depth analysis on the subject, and perhaps, this has inhibited research in the field. Yet, there is substantial material available, for instance, in the report of the Fuel Policy Committee (FPC), which is valuable and relevant for energy researchers. For this reason, there has been an attempt to provide a large number of tables and as much data as possible from existing sources to substantiate comments and analysis.

This effort would not have led to fruition were it not for the help and encouragement given to the author by various individuals throughout the duration of the venture. I am especially grateful to N. P. Sen, principal of the Administrative Staff College of India, for his valuable encouragement and support and A. K. Das Gupta, of the same college, for his help in locating and arranging various materials required for this study (and for satisfying my whims). I am full of gratitude and admiration for my secretary, Samuel Abraham, for an immaculate job of typing performed in record time. The help provided by Ida D'Souza and Irene Paes is gratefully acknowledged. I would also like to express my thanks to the following for permission to reproduce materials as indicated: Oxford University Press, London, for Table 3.4 on p. 10 from Forest Energy and Economic Development, by D. E. Earl; the Ford Foundation for Table 1-2 on p. 8 from Energy and Agriculture in the Third World, by Arjun Makhijani, published by Ballinger Publishing Company, Cambridge, Massachusetts; Addison-Wesley Publishing Co., for Tables 15.6-1 on p. 429, 15.6-2 on p. 430, 15.6-3 on p. 431, and Figure 14.3-1 on p. 337 from Energy, vol. 2: Non-Nuclear Technologies, by S. S. Penner and L. Icerman; IPC Science & Technology Press, Guidford (U.K.), for Figure 1 on p. 4 from "US Energy Policy Evaluation: Some Analytical Approaches," by Dilip R. Limaye and John R. Sharko, in Energy Policy 2, no. 1 (March 1974); and Macmillan Company of India for Table 4.2 on p. 50 from Second India Studies: Energy, by Kirit Parikh.

I am guilty of having committed at least one crime while writing this book, namely, that of using bonded labor, even though the practice has been legislated against in India. In acknowledgment and atonement for this sin, I must express my thanks to my wife, Saroj, for giving me the initial idea of writing this book, and for help at every stage, and, in particular, for art work on the diagrams; my mother-in-law, Mrs. Basant Puri, for checking all the tables; my daughter, Rashmi, a budding economist, for proofreading and helping with material in Chapter 2; and to daughters Shona and Moneesha for not bringing the roof down while everyone else was working. My final thanks to my parents, Dr. and Mrs. A. R. Pachauri, and my sister-in-law, Mrs. Sarita Pachauri, for their support and encouragement.

CONTENTS

	Page
PREFACE	vi
LIST OF TABLES AND FIGURES	x

Chapter

1	INTRODUCTION	1
	Notes	16
2	INDIAN ECONOMY	17
	Review of Indian Planning	18
	First Five-Year Plan	18
	Second Five-Year Plan	19
	Third Five-Year Plan	20
	Growth Perspectives Up to 1971	21
	The Indian Economy in the 1970s	23
	Past Growth and Future Prospects	24
	Socioeconomic Development	41
	Notes	46
3	ENERGY DEMAND	47
	Determinants of Demand for Various Energy Sources	49
	Forecasts of Demand for Different Energy Sources	57
	Notes	68
4	ENERGY RESOURCES	69
	Petroleum	69
	Coal	76
	Hydroelectric Resources	87
	Nuclear Resources	90
	Solar Energy Resources	99
	Biogas Resources	100
	Wind Power Resources	103

Chapter		Page
	Forest Resources	105
	Notes	105
5	TECHNOLOGICAL ALTERNATIVES	107
	Technological Alternatives to Energy from Coal	107
	Hydroelectric Power	111
	Electric Power Generation	113
	Noncommercial and Nonconventional Sources of Energy	117
	Solar Energy	117
	Biogas Energy	127
	Wind Energy	133
	Forest Energy	138
	Organizational Aspects	143
	Notes	144
6	POWER SECTOR	145
	Managerial and Administrative Problems in the Power Sector	150
	The Power Equipment Industry in India	166
	Notes	178
7	CONCLUSIONS	179
ABOUT THE AUTHOR		187

LIST OF TABLES AND FIGURES

Table		Page
1.1	Indexes of Energy Consumption and Selected Economic Indexes, 1960/61, 1961/62, and 1965/66–1970/71	5
1.2	Per Capita Gross National Product and Energy Consumption for Selected Countries	6
1.3	Employment and Labor Productivity in Rice Production	8
1.4	Coal Replacement and Coal Equivalent of Different Fuels	11
1.5	Consumption of Commercial Energy in Various Sectors, 1953/54, 1960/61, 1965/66, and 1970/71	12
1.6	Rate of Growth of Commercial Energy Consumption in Various Sectors, 1953/54–1960/61, 1953/54–1970/71, 1960/61–1965/66, and 1965/66–1970/71	13
1.7	Consumption of Commercial Energy in Various Sectors, 1953/54 and 1970/71	14
1.8	Consumption of Noncommercial Energy in the Domestic Sector, 1960/61–1970/71	15
2.1	Growth Rate in Gross Domestic Product at Factor Cost, 1961/62–1973/74	25
2.2	Summary Profiles of Levels of Agricultural Development at the District Level, 1970/71–1972/73	26
2.3	Summary Profile of Growth of Agricultural Development at the District Level, from 1962/63–1964/65 to 1970/71–1972/73	27
2.4	Systematic Hydrogeological Survey, as of January 1, 1975	29
2.5	Life of Known Reserves at 1988/89 Consumption Levels	32

Table		Page
2.6	Projected Sectoral Rate of Growth in Gross Value of Output and Gross Value Added at Factor Cost for the Fifth Five-Year Plan and Sectoral Composition of Gross Value Added in 1973/74 and 1978/79	33
2.7	Actual Physical Output Levels, 1973/74, and Projections of Physical Output Levels, 1978/79	34
2.8	Domestic Savings by Generating Sectors	35
2.9	Domestic Savings, by Sector of Origin, 1973/74 and 1978/79	36
2.10	Export Projections for the Fifth Five-Year Plan Period	38
2.11	Import Projections for the Fifth Five-Year Plan Period	39
2.12	Outlays for Important Industries, 1974/75-1978/79	40
2.13	Revised Fifth Five-Year Plan Outlay on Railways	42
2.14	Revised Outlays for Central Programs	43
3.1	Consumption of Petroleum Products by End Use	52
3.2	First Estimates of Coal Requirements, End-Use Method, 1978/79, 1983/84, and 1990/91	59
3.3	First Estimates of Demand for Oil Products, End-Use Method, 1978/79, 1983/84, and 1990/91	60
3.4	Revised Estimates of Demand for Electricity, End-Use Method, 1978/79, 1983/84, and 1990/91	62
3.5	Total Electrical Energy Requirements, 1978/79-1983/84 and 1990/91	63
3.6	Estimates of Demand for Noncommercial Fuels, 1978/79, 1983/84, and 1990/91	65

Table		Page
3.7	Estimated Requirements of Fuels, 1978/79, 1983/84, 1990/91, and 2000/2001, and Actual Consumption, 1970/71	67
4.1	Oil and Natural Gas Commission's Output of Crude Oil, 1967/68–1974/75	74
4.2	Sedimentary Areas, by Category	77
4.3	Reserves of Coal	80
4.4	Regional Availability of Coal Reserves, by Category	82
4.5	Coal Washeries, 1951–70	84
4.6	Growth of Coal Production, 1900 to 1973/74	85
4.7	Distribution of Coal Mines, by Size, 1970/71	86
4.8	Possible Levels of Production of Coal from Different Coalfields up to 1990/91	88
4.9	Distribution of Power Potential, by State	89
4.10	Hydel Energy Potentials of Various Regions	90
4.11	Plant Investment Costs in the Agro-Industrial Complex	93
4.12	Details of the Agricultural Side of the Agro-Industrial Complex	93
4.13	Overall Capital Costs of the Agro-Industrial Complex, 1976	94
4.14	Distribution of Biogas Plants, by States, as of March 31, 1975	102
4.15	Distribution of the Forest Area, by State, 1969/70	104
5.1	Relative Wind Power System Costs, Power Capacities, and Costs per Unit of Capacity as Functions of the Design Wind Speed	136

Table		Page
5.2	Relative Specific Output of Wind Power Systems as a Function of the Design Wind Speed for Sites with Different Annual Mean Wind Speeds	137
5.3	Relative Cost per Unit of Electrical Energy Output for Different Wind Power Systems	138
5.4	Comparison between the Percentage of Moisture Content and Calorific Value of Wood, Determined on a Dry Weight and Wet Weight Basis	140
6.1	Electricity Generation, 1960/61–1990/91	146
6.2	Per Capita Consumption of Electricity, 1960/61–1990/91	146
6.3	Fifth Five-Year Plan: Power—Financial Outlay	147
6.4	Breakdown of the Installed Capacity at the End of the Fourth Five-Year Plan and Fifth Five-Year Plan, by Region, by Type of Plant, as of March 31, 1974 and March 31, 1979	148
6.5	Forecasts of Demand for Electrical Energy by Three Alternative Methods	150
6.6	Estimated Models for Electricity Demand in the Domestic Sector of Andhra Pradesh	151
6.7	Estimated Models for Electricity Demand in the Commercial Sector of Andhra Pradesh	152
6.8	Estimated Models for Electricity Demand in the Industrial High Tension Sector of Andhra Pradesh	153
6.9	Estimated Models for Electricity Demand in the Agricultural Sector (Low Tension and High Tension) of Andhra Pradesh	154
6.10	Arrears of Interest Outstanding from State Electricity Boards on Loans from State Governments, as of End of 1973/74	156

Table		Page
6.11	Physical Achievements of Rural Electrification Corporation Schemes as of March 31, 1976	162
6.12	Groundwater Resources	164
6.13	Electrified Villages	165
6.14	Statement of Line Losses, Length of Lines, Number of Agricultural Pump Sets, as of April 1972	167
6.15	Financial Returns of State Electricity Boards, 1971/72–1973/74	168
6.16	Addition to Installed Capacity in the Country's Hydroelectric, Thermal, and Nuclear Power Plants, 1971/72–1975/76	174
6.17	BHEL's Order Position, as of March 31, 1976	175
6.18	BHEL's Physical Output, Actuals, 1969/70–1975/76	176

Figure		
1.1	Relationship between per Capita Energy Consumption and per Capita GNP	3
1.2	Relationship of Energy Consumption with GNP in the United States	4
4.1	Drilling Performance	78
4.2	Number of Wells Worked Over by Work Over Rigs	79
4.3	Strategy for Nuclear Power in India	95
5.1	Combined Cycle Generation of Power	116
5.2	A Typical Solar Collector	119
5.3	Vapor-Pulse Engine	120
5.4	Solar Space and Water-Heating System for a Typical House	121

Figure		Page
5.5	The Ocean Thermal Energy Conversion Cycle for Energy Production	126
5.6	Layering of By-products in the Digester	129
5.7	Gas Production and Temperature	130
5.8	Comparison of Gas Production Rates at 60° and 95°F	131
6.1	Financial Results of the Rural Electrification Corporation	161

Energy and Economic Development in India

CHAPTER

1

INTRODUCTION

Industrial societies today are characterized by consumption patterns demanding large quantities of energy and fuel. Modern technology and living styles have evolved on the expectation of uninterrupted supplies of conventional energy forms. This expectation was disturbed by the waves that followed the Arab oil embargo of 1973 and the concurrent price increases in oil brought about by the Organization of Petroleum Exporting Countries (OPEC) cartel. The ripples from this major international development have still not died down and are likely to be felt for years to come. Whereas concerns were raised in the developed world with respect to maintenance of the high standards of living that had been attained in the twentieth century, the oil-poor developing nations of the world were confronted by questions of their very survival. Having weathered the storm at its height, positive plans and programs are now being evolved to solve the energy crisis (the commotion in international oil markets was merely a symptom and not the cause of the crisis).

Before the advent of the "green revolution" in the developing world, increases in the consumption of energy were associated primarily with activities in transport, mining, and manufacturing. Since the adoption of new methods of large-scale farming, the agricultural sector in most developing countries has also become a growing consumer of energy, both directly and indirectly. When we add to these sectors the demands for energy for residential consumption by a large population, as in the case of India, energy acquires a priority over many other resources essential for economic well-being and development. The relatively low prices of oil and its abundant availability in the past few decades gave rise to the evolution and adoption of technologies that depended on petroleum. Hence, even though rapid increases in power and coal consumption accompany national economic growth, the rate of increase of petroleum consumption has been higher in developing nations in the past three decades. In India, for instance, the Fuel Policy Committee (FPC) indicated that by 1971, approximately one-half of the total commercial energy consumed was in the form of petroleum.[1]

The relationship between economic growth and energy consumption is brought out in Figure 1.1, in which gross national product (GNP) per capita for selected countries is plotted against their respective per capita energy consumption levels. The pattern observed clearly establishes a strong correlation between economic development and energy consumption. The case of China, which stands out as an exception, indicates a less energy-intensive pattern of development than that established in other countries. The Indian model of economic development, however, is somewhat similar to those of Western countries, and hence future consumption of energy is likely to be related to economic growth in a manner similar to that indicated in the diagram.

The dynamic relationship between energy consumption and GNP is brought out more sharply in Figure 1.2, which shows a plot of these variables, as well as the ratio between them, for the period 1947-70 for the United States. This figure indicates that there is a tendency for the ratio of energy consumption to GNP to decrease with higher levels of economic growth. Annual energy consumption data for India have not been compiled by official sources, except for some selected fiscal years.* These are shown for selected years, along with corresponding figures for net national product, industrial production, population, and so forth, in Table 1.1. In this table, the ratio of total energy consumption and net national product is shown for India; a rising trend is exhibited over part of the period covered and then there is a relative decline. With a further broadening of the industrial base and transportation facilities and improved utilization of existing resources, this ratio is likely to decrease with overall economic growth.

Published international statistics on energy consumption and growth can, however, give a misleading impression about the stage of economic progress attained by a country, mainly because the use of noncommercial forms of energy, such as firewood, vegetable and animal wastes, and so forth, are excluded from such data. In the case of India, for instance, it is estimated that energy from noncommercial sources takes up approximately 50 percent of the total energy consumed in the country. D. E. Earl has compiled a table showing the proportion of energy supplied from fuel wood, along with other indicators, for selected countries.[2] This has been reproduced in Table

---

*The fiscal year in India runs from April 1 to March 31.

FIGURE 1.1

Relationship between Per Capita Energy Consumption and Per Capita GNP

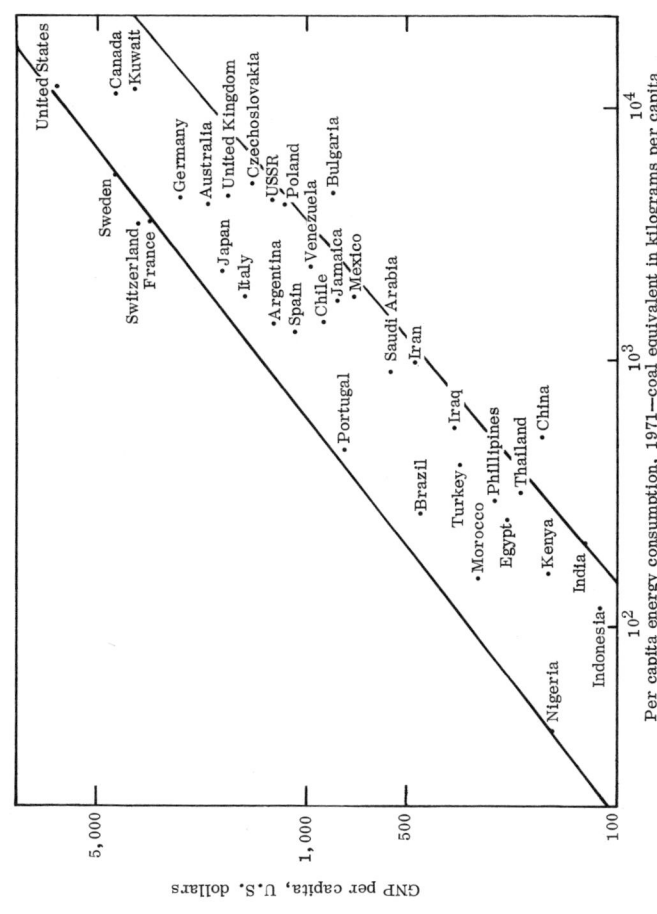

Source: K. Krishna Prasad and A.K.N. Reddy, "Technological Alternatives and the Indian Energy Crisis," Paper presented at the National Seminar on Energy, Hyderabad, March 1976.

FIGURE 1.2

Relationship of Energy Consumption with GNP in the United States

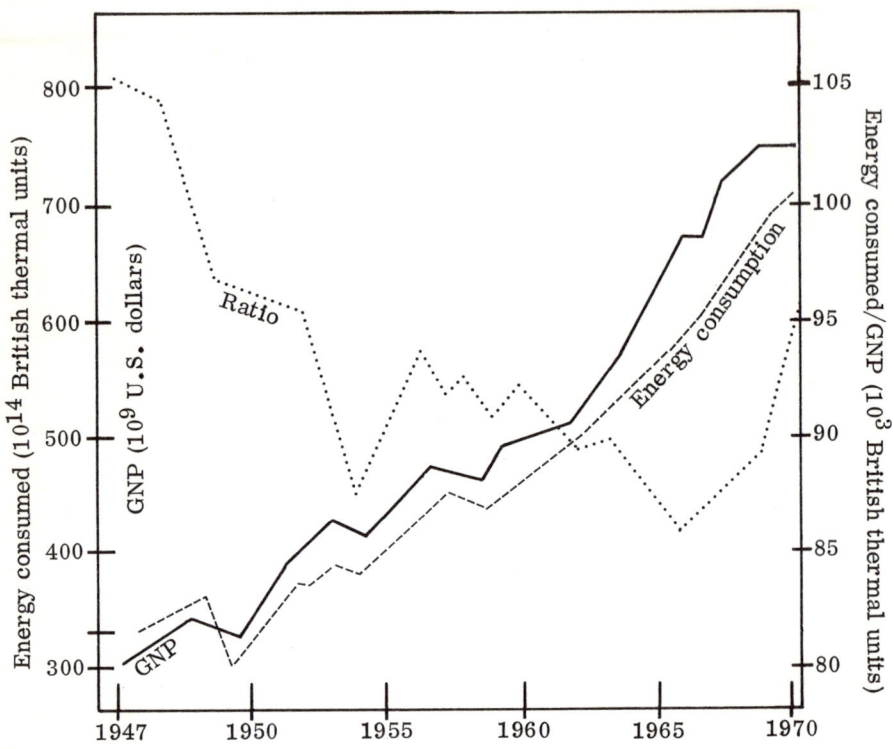

Source: Dilip R. Limaye and John R. Sharko, "U.S. Energy Policy Evaluation: Some Analytical Approaches," in Energy Policy, vol. 2, no. 1 (Guidford, U.K.: IPC Science and Technology Press, March 1974).

1.2. From this table, it can be seen that forests are a major source of energy in many countries, particularly those of East Africa, but they supply very little energy in Western Europe and North America.

The consumption of noncommercial sources of energy in developing countries is concentrated largely in the rural sector, which is engaged predominantly in agricultural activities. A change from the use of animal waste and firewood as sources of energy to commercial sources would require extension of markets into, and increased in-

TABLE 1.1

Indexes of Energy Consumption and Selected Economic Indexes, 1960/61, 1961/62, and 1965/66–1970/71 (1960/61 = 100)

Factor	1960/61	1961/62	1965/66	1966/67	1967/68	1968/69	1969/70	1970/71
Net national product	100.0	103.4	113.5	114.7	125.0	129.2	136.2	142.0
Industrial production	101.7	111.2	150.3	152.8	153.6	163.6	175.1	181.5
Net domestic product in mining and manufacturing	100.0	109.0	139.0	140.0	143.0	149.0	157.0	160.0
Population	100.0	102.1	111.3	113.8	116.3	118.9	121.6	124.3
Total commercial energy consumption (excluding domestic sector consumption) in coal replacement terms	100.0	110.8	150.0	157.9	169.3	181.1	196.5	201.3
Total energy consumption (including noncommercial energy) in coal replacement terms	100.0	105.1	124.7	129.2	134.8	140.6	147.7	151.2
Ratio of total energy consumption to net national product	1:0000	1:0164	1:0986	1:1264	1:0784	1:0882	1:0844	1:0648

Note: Indexes constructed with reference to population figures in 1961 and 1971. Population in intermediate years has been projected by suitable interpolation.

Sources: Government of India, Economic Survey—1973–74; and Government of India, Central Statistical Organization, Monthly Statistics of Industrial Production, and Estimates of National Product, all months from April 1960 to March 1971 (New Delhi: Government of India Press, 1974).

TABLE 1.2

Per Capita Gross National Product and Energy Consumption for Selected Countries*

Country	GNP per Capita (U.S. dollars)	Consumption per Capita Fuel Wood (m³)	Energy Consumption per Capita (kilograms of coal equivalent)		Proportion Total Energy Supplied by Fuel Wood (percent)
			Forest Only	Total	
Malawi	80	0.77	335	376	89.1
Nepal	80	0.57	248	259	95.8
Tanzania	100	2.30	999	1,042	96.0
India	110	0.19	83	274	30.3
Sri Lanka	110	0.31	135	291	46.4
Guinea	120	0.50	217	314	69.1
Nigeria	120	1.00	435	480	90.6
Malagasy	130	0.52	240	304	78.9
Uganda	130	1.07	478	531	90.2
Kenya	150	0.69	299	447	66.9
Rhodesia	280	0.63	274	838	32.7
Algeria	300	0.02	9	479	1.9
Ivory Coast	310	1.01	438	618	70.9
Zambia	400	0.90	391	900	43.4
Brazil	420	1.60	695	1,176	59.1
Cuba	530	0.20	87	1,140	7.6
Chile	720	0.31	135	1,345	10.4
South Africa	760	0.04	17	2,763	0.6
Venezuela	980	0.63	274	2,427	11.3
Greece	1,090	0.25	109	1,259	8.7
Italy	1,760	0.14	61	2,492	2.4
Libya	1,770	0.20	87	569	15.3
USSR	1,790	0.36	157	4,356	3.6
United Kingdom	2,270	0.01	4	5,143	0.1
Finland	2,390	1.63	709	4,859	14.6
Belgium	2,720	0.20	9	5,438	0.2
West Germany	2,930	0.03	13	4,836	0.3
France	3,100	0.12	52	3,570	1.5
Canada	3,700	0.20	87	8,881	1.0
Sweden	4,040	0.41	178	5,946	3.0
United States	4,760	0.10	43	10,817	0.4

*Based on data from: United Nations (1973), International Bank for Reconstruction and Development (1972), Food and Agriculture Organization (1972).

Source: D. E. Earl, Forest Energy and Economic Development (London: Oxford University Press, 1975), p. 10.

INTRODUCTION

comes within, rural areas. Such developments have been found to take place in regions with modernized agricultural methods. The aggregate effect of modernization in agriculture, therefore, is not necessarily a corresponding increase in overall energy consumption; to a large extent, the aggregate effect is a shift to commercial sources from noncommercial sources.

Most energy consumed in the rural sector of developing countries is used in inefficient ways to yield low levels of useful output, making comparisons of per capita energy usage between developing and developed countries somewhat meaningless. Further, to a substantial degree, human and animal power are substitutes for other energy sources, and the availability in abundance of these types of power at relatively low prices leads to adoption of labor- and animal-intensive production technologies. The marginal product of labor with such technologies remains low, nevertheless, and can be increased only with corresponding increases in capital and in conventional energy inputs.

Apart from increases in consumption of fertilizers, pesticides, and high-yielding varieties of seeds, agricultural development (in the conventional form) involves mechanization of certain agricultural operations, such as threshing, harvesting, tilling, and plowing, which may or may not reduce peak demands on the rural labor market. Another agricultural input requiring larger energy usage is irrigation. Improved irrigation facilities do not result in displacement of labor, as some of the other inputs of modernized agriculture may. On the contrary, increased irrigation permits cultivation of larger acreages and multiple cropping, leading to greater employment opportunities for rural labor. The scope for increased irrigation and fertilizer usage in India can be seen from Table 1.3, which shows the position of India in relation to three other Asian countries.

In developing countries, the demand for various inputs in different organized sectors is sometimes more predictable than in an industrialized and developed economic system. This is due to the fact that the mix of technologies adopted for production and the constant upgrading of technology are the results of careful planning and generally move on predictable lines, drawing on existing available technologies rather than on the uncertainties of innovation. Developing countries, such as India, have large inequalities in income distribution, which exist not only between the urban and rural sectors but within the rural sector itself. For a country that has 75 percent of its population in rural areas, changes in income distribution in agricultural occupations would have a very large impact on the technologies and scales of productive activities in agriculture. It is not unusual to see coexisting in Indian villages modernized agricultural techniques of the Western model and age-old practices in farming (using largely animal or human energy).

TABLE 1.3

Employment and Labor Productivity in Rice Production

Country	Number of Workers per 100 Hectares (1965)	Labor Productivity, Kilograms of Rice per Worker per Crop[a]	Land Productivity, Kilograms of Rice per Hectare (1969)	Percentage of Rice-Producing Land Irrigated[b]	Nitrogen Fertilizer Application, Kilograms per Hectare[c] (1970) (approximate)
Japan	215	2,600	5,600	100	150
China	180–200	1,500	3,000	70	50[d]
Taiwan	195	2,050	4,000	100	100
India	90	1,550	1,400	40	10–20

[a]In Japan and Taiwan, multiple cropping (three or four crops per year) is common, so that the annual productivity of the worker is several times his productivity per crop. Multiple cropping is less common in China and India.

[b]The percentage of irrigated area under rice cultivation is generally higher than national averages for all crops, because rice-producing areas are irrigated in preference to other crops.

[c]Includes the nitrogen content of organic fertilizers, which supply about 60 percent of the nitrogen.

[d]Nitrogen application is averaged over the entire area planted with rice. In practice, chemical nitrogen is applied to only part of the area. The average, therefore, indicates the extent of use rather than the typical practice in application of nitrogen per hectare.

Source: Arjun Makhijani, Energy and Agriculture in the Third World (Cambridge, Mass.: Ballinger Publishing Company, 1975), p. 8.

INTRODUCTION

The redistribution of land and implementation of land reforms can, therefore, make a large impact on the type of farming practices adopted in the years to come. With a breakup of larger farms enforced by land reform legislation, it is likely that a shift toward more labor-intensive agricultural techniques will take place in some parts of the country. It is unlikely, however, that the recent trend of using high-yielding varieties of seed, requiring greater inputs of water and fertilizers, will be reversed.

Land reforms will have a large impact on internal migration as well. Absorption of surplus labor in agricultural activities within the rural sector will reduce the outflow of population into urban and semi-urban areas. In the long run, this could affect the demand for energy for transportation purposes.

Similar implications exist for the type of industrial activity that the country will adopt in the future. For instance, conventional production of fertilizers and electricity are known to provide one direct job for U.S. $20,000 of capital invested. If this were done using small-scale industries for production of electricity and fertilizers, the ratio would come down to around one job for U.S. $2,500. Whereas figures of this type may indicate that large-sized plants and production from them are not in the interests of a labor surplus economy, the solution is not all that simple. The evaluation and dispersion of technology of the small-scale type is expensive and hampered by various constraints and barriers, such as illiteracy in rural areas. Further, the managerial and organizational inputs required for efficient running of widely dispersed small-scale units pose a challenge that cannot be met easily in the near future.

The problem of adopting proper energy policies in developing countries is further compounded by the difficulties in predicting population and national income growth rates. Some predictions indicate that the population of India, by the turn of the century, will be over 1 billion. If the ordinary energy needs of such a large population are to be met, the problem posed is staggering. The scale of investments necessary in the energy sector would be so large that it is difficult to foresee surpluses being generated in the economy to the scale that would permit investments in this sector in addition to meeting the needs of other sectors of the economy. Yet, energy plays a vital role in generating economic growth. It must be noted that consumption of energy both in the household and in industrial activities is a substitute for the inputs of labor and capital. Therefore, surpluses can be diverted either through more efficient use of energy or through the substitution of greater energy consumption in order to free capital and labor for other production activities.

The lack of commercial energy, such as electricity, has been acting as a barrier to development of various kinds. For instance, rural electrification would lead to a host of other opportunities that rural areas would otherwise be denied access to. The social benefits of rural electrification have led most countries to adopt positive programs that do not rely on generation of revenues from private consumers for meeting the cost of electrification. The Indian government, through a central Rural Electrification Corporation (REC) and the operation of state-controlled electricity boards lays down specific targets for rural electrification as part of the five-year plans that are launched by the government.

Different forms of energy generally have some degree of substitution possible between different consumption activities for which they are employed. When aggregate quantities and values have to be considered without distinction between the form of energy used, a common measure has to be used. This is generally done by converting quantities in different forms of energy to one common equivalent. For instance, coal-equivalent metric ton is the most commonly used unit of measure in various international and national studies.* Since oil has, to a large extent, replaced coal as the most commonly used source of energy, oil-equivalent metric ton is being used more frequently than before, but the United Nations still uses coal-equivalent metric ton as the unit of measure for statistics compiled internationally. Again, there are variations in the values of kilocalories per kilogram for different varieties of coal used for comparison. The calorific value of Indian coal varies from 6,700 kilocalories per kilogram in the case of certain selected grades to approximately 4,000 kilocalories per kilogram for other grades. Official statistics in India generally base their comparisons on a value of 5,000 kilocalories per kilogram for Indian coal and a value of 10,000 kilocalories per kilogram for oil. Generally speaking, therefore, one metric ton of oil is expressed as being equivalent to two metric tons of coal.

Another measure that has been used in India as well as other countries is the coal-replacement value, which indicates the amount of coal that would be required to substitute the fuel being considered, taking into account the manner in which the substitution would take place. For instance, whereas the use of one metric ton of fuel oil for producing heat in industrial applications could be substituted by the use of two metric tons of coal, the use of one metric ton of high-speed diesel oil (HSDO) in railway traction would require nine metric tons of coal. This is due to differences in functioning of diesel locomotives and steam locomotives. One metric ton of fuel oil will, therefore, be considered as being measured as two coal-replacement metric tons, while one

---

*This unit expresses the heat content (kilocalories) of different fuels in terms of heat content of an average metric ton of indigenous coal.

TABLE 1.4

Coal Replacement and Coal Equivalent of Different Fuels

Fuel	Unit	Coal Equivalent in mmt	Coal Replacement in mmt
Coal products			
Coal (coking 6,640 kilocalories per kilogram; noncoking coal used in steam generation 5,000 kilocalories per kilogram)	1 mmt	1.00	1.00
Hard coke	1 mmt	1.30	1.30
Soft coke	1 mmt	1.50	1.50
Firewood (47,500 kilocalories per kilogram)	1 mmt	0.95	0.95
Charcoal (6,900 kilocalories per kilogram)	1 mmt	1.00	1.00
Oil products (10,000 kilocalories per kilogram)			
Black products (fuel oil, furnace oil, refinery fuel, low-speed/high-speed, heavy high-speed)	1 mmt	2.00	2.00
Kerosene and liquid petroleum gas	1 mmt	2.00	8.30
High-speed diesel oil and light diesel oil	1 mmt	2.00	9.00
Motor spirit and jet fuel	1 mmt	2.00	7.50
Natural gas (9,000 kilocalories per kilogram)	$10^9 m^3$	1.80	3.60
Electricity	$10^9 m^3$	1.00	1.00

Note: mmt = million metric tons.

Source: Report of the Fuel Policy Committee (New Delhi: Controller of Publications, 1974), Table 2.1, p. 5.

metric ton of HSDO will be reckoned as nine coal-replacement metric tons. Since the coal-replacement measure has been prevalent in India for a number of years, official statistics generally contain figures in terms of coal equivalent as well as coal replacement. Table 1.4 gives the value of different commercial fuels used in India in terms of coal equivalent and coal replacement.

TABLE 1.5

Consumption of Commercial Energy in Various Sectors,
1953/54, 1960/61, 1965/66, and 1970/71[a]
(millions of metric tons)

Sector	Consumption of Commercial Energy			
	1953/54	1960/61	1965/66	1970/71
Mining and manufacturing	22.4	39.7	60.8	76.3
	(37.3)[b]	(39.2)	(41.4)	(38.7)
Transportation	21.5	34.2	49.7	64.6
	(35.8)	(33.8)	(33.8)	(32.7)
Agriculture	1.8	3.6	6.3	9.1
	(3.0)	(3.5)	(4.3)	(4.6)
Domestic	12.7	20.8	26.5	35.5
	(2.0)	(20.6)	(18.0)	(18.0)
Government, commercial, and other	1.7	2.9	3.7	11.8
	(2.8)	(2.9)	(2.5)	(6.0)
Total	60.1	101.2	147.0	197.2
	(100.0)	(100.0)	(100.0)	(100.0)

[a]Calculated with reference to coal-replacement terms.
[b]Numbers in parentheses represent percentages of the total for each fiscal year.

Source: *Report of the Fuel Policy Committee* (New Delhi: Controller of Publications, 1974), Table 2.5, p. 6.

Economic activities, as mentioned above, bring about changes in fuel consumption patterns. The growth of activities in industry, transportation, and agriculture lead to rapid increase in the consumption of energy. The mix of different fuels depends on the relative price between different energy forms and the technologies that evolve in these areas. In the aggregate, the share of different sectors in overall consumption of energy has been changing for the period 1953-71 (data for these years have been compiled and published). The consumption in different sectors in this period, as well as rates of growth in certain sectors, are shown in Tables 1.5 and 1.6, respectively. If the consumption of different forms of energy is examined for the same period, it is observed that the percentage share of coal in all the sectors has been declining over the last two decades. An active program by the government to promote consumption of coal—which

INTRODUCTION

has been reinforced by increasing prices of oil and electricity—may result only in slowing down this trend without necessarily reversing it. The percentage shares of the consumption of these three forms of energy (consumed sectorally) are shown in Table 1.7. It is shown that the rate of growth of oil consumption has been highest in the transport sector, whereas the rates of growth of electricity have been highest in the mining and manufacturing, domestic, and agricultural sectors.

Energy problems in India, as in other developing countries, are not confined only to commercial energy supply and demand. Since the consumption of noncommercial energy is unusually large, as mentioned before, it requires analysis and study in as much depth as that done for organized commercial consumption. Apart from the consumption of forest fuels, cow dung, and vegetable wastes, another form of energy that is used in large measure in rural areas is animal energy, which usually takes the form of bullocks, buffalo, camels, and so forth, mainly for the purpose of tilling, drawing water, or transport-

TABLE 1.6

Rate of Growth of Commercial Energy Consumption in Various Sectors, 1953/54-1960/61, 1953/54-1970/71, 1960/61-1965/66, and 1965/66-1970/71*

Sector	Percentage Average Annual Rate of Growth of Commercial Energy per Annum			
	1953/54–1970/71	1953/54–1960/61	1960/61–1965/66	1965/66–1970/71
Mining and manufacturing	7.5	8.5	8.9	4.6
Transportation	6.9	6.9	7.8	5.4
Domestic	6.2	7.3	4.9	6.0
Agriculture	9.9	10.0	12.2	7.5
Government, commercial, and other	12.1	7.9	4.9	26.2
All sectors	7.2	7.7	7.6	6.1

*Calculated with reference to coal-replacement terms.

Source: Report of the Fuel Policy Committee (New Delhi: Controller of Publications, 1974), Table 2.6, p. 7.

TABLE 1.7

Consumption of Commercial Energy in Various Sectors,
1953/54 and 1970/71

Sector and Fuel	1953/54 Mtcr[a]	Percent-age	1970/71 Mtcr[a]	Percent-age
Mining and manufacturing	22.45	100.00	76.32	100.00
Coal[b]	13.80	61.47	31.07	40.71
Oil[c]	3.65	16.26	10.90	14.28
Electricity	5.00	22.27	34.35	45.01
Transport	21.46	100.00	64.57	100.00
Coal[b]	12.10	56.38	15.91	26.64
Oil[c]	8.76	40.82	47.23	73.15
Electricity	0.60	2.80	1.43	2.21
Domestic	12.69	100.00	35.48	100.00
Coal[b]	2.20	17.34	4.07	11.47
Oil[c]	9.79	77.15	27.58	77.73
Electricity	0.70	5.51	3.85	10.80
Agriculture	1.81	100.00	9.05	100.00
Coal[b]	—	—	—	—
Oil[c]	1.61	88.95	4.51	49.83
Electricity	0.20	11.05	4.54	50.17
Government, commercial, and other	1.70	100.00	11.77	100.00
Coal[b]	0.60	35.29	0.30	2.55
Oil[c]	—	—	—	—
Electricity	1.10	64.71	4.50	38.23

[a]Million metric tons coal replacement.
[b]Exclusive of coal used for power generation.
[c]Oil products used for energy purposes only.

Source: Report of the Fuel Policy Committee (New Delhi: Controller of Publications, 1974), Table 2.7, p. 7.

INTRODUCTION

TABLE 1.8

Consumption of Noncommercial Energy in the Domestic Sector, 1960/61–1970/71

Year	Total Noncommercial Energy	Firewood (mmtcr)*	Cow Dung (mmtcr)*	Vegetable Waste
1960/61	147.67	95.99	22.15	29.53
1961/62	149.64	97.27	22.44	29.93
1962/63	151.38	98.40	22.70	30.28
1963/64	156.22	101.54	23.43	31.25
1964/65	158.12	102.78	23.72	31.62
1965/66	163.43	106.23	24.51	32.69
1966/67	166.92	108.50	25.04	33.38
1967/68	170.87	111.07	25.63	34.17
1968/69	173.24	112.61	25.99	34.64
1969/70	175.76	114.24	26.37	35.15
1970/71	179.41	116.62	26.91	35.88

*Million metric tons coal replacement.

Source: Report of the Fuel Policy Committee (New Delhi: Controller of Publications, 1974), Table 2.9, p. 8.

ing goods and passengers. These forms of energy are generally not used for industrial applications. The FPC arrived at some tentative estimates of the consumption of noncommercial energy in the domestic sector. These are shown in Table 1.8.

The problem of consumption of noncommercial energy lies not merely in the extensive quantities that are used but also in the inefficient manner in which energy is converted to perform various functions. The efficiency of conversion associated with these noncommercial sources of energy is deplorably poor, and since most of these are consumed without any cost to the consumer, there is little incentive to adopt improved methods. According to some estimates, the indiscriminate felling of trees for firewood has resulted in the forest area in India being reduced from 25 to 23 percent of the total land mass in a period of about 30 years (since independence). The costs of such waste are extremely high, since deforestation not only promotes soil erosion but also affects rainfall.

The problems posed by consumption of energy in this manner cannot be solved in the short run, given existing institutions and politico-legal systems. Considerable research is being done, for instance, on the possibility of fast-growing forests, which would serve rural areas by providing cheaper firewood, which could be constantly replenished. The implementation of plans for large-scale introduction of biogas plants, using cow dung as the feedstock, is also being widely adopted and advocated by a number of individuals and agencies. However, solutions of this type in a country the size of India can only make an impact after a long period of time. In the interim, with the resources, administrative infrastructure, and knowledge available, quicker solutions can perhaps be found in the sphere of commercial consumption of energy. However, this does not mean that our priorities should be diverted toward finding solutions to our energy problems in more efficient use of commercial energy to the exclusion of noncommercial forms. While research and development and trials proceed for the noncommercial sources, it would be logically sound to implement (at least) those solutions in the commercial sector that appear within reach at present.

In the following chapters, we will present a survey of the Indian economic scene, offer some tentative forecasts of demand for the future, and examine the determinants of demand for energy, the various technological options available, and policy implications of various measures that can be adopted in dealing with the energy problem that is facing India. The implications of the analysis hold true not only for the Indian scene but have important lessons for other developing countries, particularly those that are net importers of crude oil.

NOTES

1. Fuel Policy Committee, Report (New Delhi: Controller of Publications, 1975), Table 2.3, p. 6.
2. D. E. Earl, Forest Energy and Economic Development (London: Oxford University Press, 1975), p. 10.

CHAPTER 2

INDIAN ECONOMY

Since the independence of India, researchers have been directing their attention to the country's record of economic development. Most studies assign a poor ranking to India in comparison with some other developing nations, which appear to have made rapid strides in recent decades. Inevitably, scholars draw a comparison between China and India, and more often than not, such studies are preceded by actual visits to either country (or both), which may lead to a biased recording of impressions, unsubstantiated by facts or analysis.

India is one of the poorest countries in the world in terms of per capita income. Even though the efforts of its people have borne fruit in a large range of activities, the acute problems of poverty affecting more than half of its (ever expanding) population have eroded the value of its achievements. Those who find satisfaction in its performance do so by assigning importance to its building of a democratic structure based on universal suffrage. Indeed, the general elections of 1977, which installed a new government in office after almost 30 years of rule by the Congress Party, did, in considerable measure, demonstrate the existence of deep roots of democratic behavior on the part of the Indian people.

Progress must, nevertheless, be measured by accepted indicators of economic and social gains. It would, therefore, be useful to survey economic developments in India, particularly since independence, when a new political and social order was established. An insight into various facets of Indian economic development would also help in understanding the role of energy in providing necessary inputs for change and economic well-being of the nation.

REVIEW OF INDIAN PLANNING

The Indian economic experience has been characterized by the formulation and implementation of a series of five-year plans, of which the fifth is currently being implemented. The emphasis on planned growth has varied from time to time in India's short history as an independent nation, and there have been periods when planning for economic development was either totally or partially abandoned. It is nevertheless useful to view the changes in economic activity in terms of five-year plan periods in order to view the developments from a longer (and more logical) perspective.

First Five-Year Plan

The First Five-Year Plan (1951/52-1955/56) can be regarded as having been a successful exercise in centralizing planning. A large number of targets were surpassed in many activities. For instance, agricultural production targets were not only reached but actually exceeded. National income in the five-year period rose by approximately 18 percent, real per capita income by 11 percent, and per capita consumption by 9 percent. The rate of investment in the economy increased from about 5 percent in 1950/51 to approximately 7.3 percent in 1955/56. Food grains production increased by approximately 20 percent, and the output of cotton, which was the main input for the textile industry, rose by 45 percent. Irrigation facilities were successfully provided for 16 million acres of land, with approximately 6 million acres being covered by major irrigation works and 10 million from minor and medium irrigation works. Growth on the industrial front was marked by an increase of the industrial production index (1950/51 equals 100) to 139, thereby achieving an annual average growth rate of about 8 percent. Consumption of cloth of different varieties per capita increased from 9.2 yards per annum to 15.5 yards per annum.

These quantitative indicators, however, are not sufficient to provide a picture of the various changes that took place in economic activity and society at large in India. A large number of institutions were created that gave the economy a certain amount of strength and stability. These included an expansion of community development projects on a large scale, village panchayats (councils), and cooperatives in order to provide a sense of participation in democracy at the grass-roots level. By the end of the plan, the cooperative movement had spread to approximately 120,000 villages, which meant that almost one-quarter of the rural population in the country had been covered by this movement. Various autonomous boards were set up during the plan period to prepare and aid programs in small-scale industries and in the promotion of cottage crafts and local enterprise. Irrigation works, rural electrification, and welfare schemes for backward and tribal areas were also launched with considerable vigor.

A major change that was brought about by the Indian government was the abolition of the <u>zamindari</u> system, which had encouraged absentee landlordism and was regarded as inhibiting the growth of agriculture. Reform of tenancy legislation—protecting the right of tenants—was another step that was taken to secure social justice for millions of cultivators, as well as to provide a push to agricultural activities.

It could be concluded that India's First Five-Year Plan was a success, especially in the fields of agriculture, irrigation, and social services. It did, nevertheless, contain a number of weaknesses, particularly since it was conceived and put together in haste and was based on scanty and unreliable statistics. It was also the country's first experiment in centralized planning and, therefore, contained certain methodological lacunae. The entire experience, however, was useful, because it established a tradition and basis for planning in the future. The First Five-Year Plan was essentially an exercise to consolidate the efforts in rehabilitation that had to be launched soon after independence because of the large-scale destruction and damage that had occurred during the riots that accompanied the partition of the country and during the World War II that preceded it. It is to the credit of the First Five-Year Plan and the central government that implemented it that shortages in the economy were removed to a large degree and that prices came down by almost 13 percent by the end of the plan period, with a corresponding decline in the cost of living index.

Second Five-Year Plan

The Second Five-Year Plan (1956/57-1960/61) was launched at a time when an optimistic mood prevailed in the country and when the economy was in fairly sound shape. The results of this plan, however, were not as satisfactory as those of the previous one. National income during this period increased by 19.5 percent, and per capita income showed a gain of only 8 percent. This smaller increase was due to the increase in population of 10.3 percent during the plan period. It is unfortunate that an extreme interpretation of Gandhian ideology on population control policy and a lack of awareness of the population problem gave rise to this increase, which also continued in subsequent years. One of the major failures of the Indian government has been a lack of awareness and action on its part to arrest the high rate of population growth in the country. Another reason for nonachievement of various targets in this plan was an assumption of a high capital output ratio, which did not materialize. P. C. Mahalanobis, who was the main author of the Second Five-Year Plan, assumed a capital output ratio of 2 to 1, whereas it actually turned out to be close to 3.86 to 1. The reason for this high figure was the greater emphasis placed on capital-intensive projects with long gestation periods as against higher investment in labor-intensive sectors of the economy.

A major effort was made in this plan to diversify agricultural production. Initially, food grains production was planned to reach 75 million metric tons, but later, the target was revised to 80 million metric tons for 1960/61. Irrigation facilities were extended to 21 million additional acres of land, with substantial increases in production and import of chemical fertilizers, seed multiplication farms, land reclamation schemes, and other extension programs. In the industrial sector, a major emphasis was on basic industries, such as iron and steel, coal, fertilizers, heavy engineering, and heavy electrical equipment. Three steel plants were set up in Durgapur, Bhilai, and Rourkela. Substantial expansion was also undertaken in production of coal and other minerals, development of India's railway network, increase of road transportation facilities, and modernization and development of major ports.

Third Five-Year Plan

The Third Five-Year Plan (1961/62-1965/66) set out to achieve an increase of 30 percent in national income, with a 70 percent increase in industry and a 30 percent increase in agricultural production. The country's efforts during this plan, however, were interrupted due to China's invasion of India in 1962 and a short war with Pakistan in 1965. Priorities in production had to be changed in favor of producing defense equipment. A further disruption was caused by failures of monsoon rains in all the years of the Third Five-Year Plan, except during 1964/65. The dependence of Indian agriculture on the monsoon makes it susceptible to abnormalities in rainfall. National income in the final year of the plan, that is, 1965/66, actually declined by 4.2 percent, although per capita income during the first four years of the plan rose by 7 percent. Wholesale prices, on the other hand, rose by 36.4 percent, and food grains production, which had reached 82 million metric tons in 1960/61, came down to 78.5 million metric tons in 1962/63, reached its highest level of 89 million metric tons in 1964/65, and fell again in 1965/66 to a level of 72 million metric tons. Overall, therefore, food production during the plan increased at an average annual rate of 2 percent, far below the targeted rate of 6 percent per annum.

The rate of growth of industrial production in the Third Five-Year Plan was a total of 28.5 percent. The main reason for the slow growth of industrial output was the emphasis on capital goods and basic intermediate goods industries as against consumer goods industries. As compared with the general index of industrial production, output in such industries as basic metals, fertilizers, heavy machinery, and generation of power grew at a reasonably rapid rate. Consumer goods industries, such as cotton textiles, showed very little increase during the third plan period. As a result, therefore, the Third Five-Year Plan had very little impact on large numbers of the Indian population; it can be characterized as a period of prolonged economic stagnation. On the employment front, it is relevant to mention that at the end of the Second Five-Year Plan, approximately 7 million people were unemployed; by the end of the Third Five-Year Plan, the labor force had increased by 17 million, but additional employment could be created only for another 14.5 million. This led to a backlog of unemployment at the beginning of the Fourth Five-Year Plan of approximately 9 to 10 million people.

The political situation in the country also led to problems in coordination between the central government and various state governments. This affected implementation of the five-year plan adversely. For instance, even though land reform legislation was assigned a high priority by the central government, reluctance on the part of the state governments did not permit it to be implemented. Further, the state governments did not accept directives from the central government to raise their resources by imposition of additional taxes. At the same time, large budgetary deficits were incurred by a number of state governments, even against the advice of the Planning Commission. It would perhaps be relevant to observe that the invasion of India by China in 1962 achieved what could have been regarded as its only objective, namely, that of thwarting India's successful efforts at economic development and all-around progress.

The brief description of the three five-year plans presented above gives an indication of the major changes and progress achieved by India in its first two decades of independence. It would be useful to review some other aspects of the overall performance of the country up to the beginning of the 1970s, when a new phase in Indian economic development began.

GROWTH PERSPECTIVES UP TO 1971

The period 1954/55–1964/65 can be taken as a sample of the two decades of economic development after independence in India. Taken as a whole, during the period 1954/55–1964-65, national income grew

at an annual average rate of 3.5 percent. This was accompanied by
a growth in population of 2.2 to 2.4 percent annually, thus leaving an
increase in per capita income of about 1.3 percent per annum. Living standards, therefore, did improve during the period, but progress
was slow because of large increases in the population. Energy problems, though important, were not acute during this period. India
continued to import the major part of its requirements of petroleum,
but even though this had an adverse effect, it did not impose a very
serious strain on the balance-of-payments position. Exports grew at
the rate of approximately 3 percent annually, but this was accompanied
by a sharp increase in imports, mainly of capital goods and selected
raw materials. This led to an increase of the trade balance from an
adverse figure of 93 crores* of rupees† in 1954 to a figure of 620
crores of rupees by 1964/65, that is, a total of 3.1 percent of the national income. At the same time, however, the ratio of net saving
to net national income rose from 8 percent in 1954/55 to 12 percent
in 1963/64, but declined to 11 percent in 1964/65. Net investment,
on the other hand, went up from 8.5 percent in 1954/55 to about 15
percent in 1963/64 and stood at 14 percent in 1964/65. The gap between savings and investment was filled by flow of capital from other
countries, which increased from 0.5 percent of national income in
1954/55 to about 3 percent in 1964/65.

In the wake of the Chinese aggression of 1962, increased outlays and production for defense purposes led to strong inflationary
pressures, which reduced savings potential and slowed down the rate
of increase in national income and employment opportunities. This
was followed by a war with Pakistan in 1965 and two years of successive drought in 1965/66 and 1966/67, which led to shortages in agricultural and industrial production. The expansion of the economy
slowed down due to a reduction in the savings to national income ratio,
which declined to 8 percent in 1966/67 as well as in 1967/68. This
led to a reduction in the rate of investment from its highest level of
15 percent in 1963/64 to 10.8 percent in 1968/69. During the year
1969/70, the savings and investment rates were 8.5 percent and 9.3
percent, respectively. A minor recovery in the economy was started
in 1967/68, which was sustained by three years of good crops, as a
result of allocations of various inputs for the agricultural sector.
In this period, the advent of the green revolution took place. This
favorable change led to reduction in imports of food and a check in
the rise of prices. An increase in agricultural production also helped

---

*One crore equals 10 million or 100 lakhs.
†One U.S. dollar equals approximately Rs 8.80 at current exchange rates.

to initiate recovery in the industrial sector. Even though growth rates in the 1960s were not as impressive as in the 1950s, population growth reached levels of 2.5 percent per annum (without any hopes of a reduction in the near future).

THE INDIAN ECONOMY IN THE 1970s

Fiscal year 1970/71 appears to have been crucial in India's economic development and political evolution. Again, there was a damaging war with Pakistan (toward the end of 1971), during which the eastern wing emerged as the new country of Bangladesh. During the major part of 1971, and up until almost the middle of 1972, India had to carry the burden of up to 10 million refugees from Bangladesh, who had crossed the borders on account of atrocities and hardships perpetrated by the Pakistani army in the eastern wing. Even though help from other countries in providing food, clothing, shelter, and medical assistance for these refugees came generously, the major burden was borne by India alone. Losses in terms of equipment and men were heavy, and this imposed an additional burden on the Indian economy. This was followed by two years of severe drought and increases in international oil prices, starting with the events of 1973. India's balance-of-payments situation slumped to a new low, with a deficit on current accounts of Rs 1,021.6 crores in 1974/75. Widespread political unrest and strikes further aggravated the economic problems of the country.

The net domestic savings expressed as a percentage of net national product declined from 14.8 percent in 1972/73 to 13.6 percent in 1973/74 and to 13.2 percent in 1974/75. A redeeming feature of this decline, however, was an increase in savings by the public sector.

In June 1975, Prime Minister Indira Gandhi imposed a state of "emergency," with stern measures against unions and opposition of various forms. This, coupled with bumper harvests and a reduction in tax evasion and smuggling, led to a temporary improvement in the economic situation in 1975/76. The decline in prices that was first observed in the second half of 1974/75 continued throughout this year. The year 1975/76 saw a reduction of the consumer price index of approximately 6 percent. Various measures that led to greater legal inflows of money and reduction in smuggling activities gave rise to a satisfactory level of foreign exchange balances.

India had a bumper crop of food grains in 1975/76, totaling approximately 120 million metric tons. This favorable trend is expected to continue in the years immediately ahead. An adequate supply of water for irrigation, large-scale use of fertilizers, and progressive

improvements in the variety of seeds used are the elements of a
strategy of agricultural growth that has been in existence during the
past three or four years. Shortfalls in growth of irrigation facilities
have caused some concern in the past year or two, and consumption
of fertilizers also has not kept pace with targets established. In fact,
during the year 1974/75, consumption of fertilizers actually declined
by 9 percent. The decline in industrial output of 0.2 percent in 1973/74
was reversed, with an increase of 2.5 percent in 1974/75. The figures for 1975/76 indicate an increase of approximately 8 percent, with
substantial increases in key industries, such as coal, iron and steel,
cement, fertilizers, power generation, aluminum, and edible oils.
Policies to liberalize government licensing for industrial expansion
are also likely to have a significant and desirable effect on the industrial sector.

An increased emphasis on agricultural production, pragmatism
in industrial policies, and buoyancy in foreign exchange reserves
spell a period of favorable economic growth for the country in the
years to come, but recent price increases indicate the reappearance
of strong inflationary forces. The single most important factor that
may inhibit the benefits of economic expansion still remains the rapid
growth of population in the country. Even though the average birthrate for the country has declined to approximately 34.5 per 1,000,
in sheer numbers, the population increases by 13 to 14 million people
every year. The strong-arm tactics by which the population control
program of the government was implemented during the period of the
recent emergency may have reduced the acceptability of population
control programs in large areas in the country. The key to economic
progress, however, lies in evolving a suitable and acceptable population control program that will reduce birthrates to near replacement
levels in the next five years or so.

PAST GROWTH AND FUTURE PROSPECTS

Problems of energy and overall development are seen in sharper
focus if one analyzes the various activities that contribute to economic
growth in a country. It would be relevant to assess growth rates in
different sectors for the immediate past. Estimates are now available up to the year 1973/74; these have been presented in Table 2.1,
which indicates growth rates in domestic product at factor cost for
different sectors covered during the period 1961/62-1973/74. It
can be observed from this table that the growth in utilities, including
electricity and the gas and water supply, has been the highest among
all the sectors detailed. In spite of this high growth rate, however,
in recent years, there have been grave shortages of electricity and

TABLE 2.1

Growth Rate in Gross Domestic Product at Factor Cost,
1961/62–1973/74*

Sector	Growth Rate (percent)
Agriculture and allied	2.07
Mining and quarrying	4.04
Manufacturing (total)	4.21
Manufacturing (registered)	4.95
Manufacturing (unregistered)	2.89
Construction	4.80
Electricity and gas and water supply	9.90
Railways	3.27
Other transport	5.16
Other services	4.35
Total	3.40

*Estimated from a semilog regression against time.

Source: Government of India, Central Statistical Organization, reproduced from Government of India, Planning Commission, Fifth Five Year Plan 1974-79 (New Delhi: Controller of Publications, 1976), Annexure 4, p. 98.

water supply all over the country. With the renewed emphasis by the present government toward providing a potable water supply for all villages in the country, growth rates for utilities, particularly with respect to the water supply, may increase at a faster rate in response to public policy and investments. The growth rate of agriculture for the period covered is only 2.07 percent. The spurt in production that took place in 1975/76 may increase this rate marginally. Present policies and efforts hold out hopes of a much higher rate of growth in agriculture than seen during this period. An analysis of agricultural development in the country, which has a direct relationship with energy demand in the agricultural sector, can be seen from Tables 2.2 and 2.3. The figures shown in these tables indicate that in the early 1970s the total value of output per hectare from agriculture amounted to Rs 1,500 per annum for only 15 percent of the total crop area. As a block of the total rural economy, this part of rural India accounted for 27.84 percent of the total output and almost 40 percent of the modern inputs of fertilizers and irrigation through pump sets and so forth.

TABLE 2.2

Summary Profiles of Levels of Agricultural Development at the District Level, 1970/71–1972/73

Gross Value of Output per Hectare (rupees)	Cumulative Percentages of Total						
	Gross Cropped Area	Aggregate Output	Consumption of Nitrogen, Phosphorous, and Potassium	Use of Tractors	Pump Sets Installed	Gross Irrigated Area	Number of Districts in India
2,500–2,799	0.70	1.83	2.37	5.39	0.83	2.24	1.06
2,000–2,499	3.04	7.18	10.60	12.89	7.82	8.27	3.56
1,500–1,999	14.48	27.84	38.93	46.81	40.68	34.08	17.73
1,000–1,499	43.90	59.46	67.24	69.90	63.40	64.25	42.91
500–999	83.96	94.20	93.79	95.88	91.56	95.75	87.94
54–499	100.00	100.00	100.00	100.00	100.00	100.00	100.00

Note: Analysis conducted for 19 main crops.

Source: Centre for Regional Development, Jawaharlal Nehru University—Perspective Planning Division, Planning Commission, Project on Regional Levels of Agricultural Development in India, reproduced from Government of India, Planning Commission, Fifth Five Year Plan, 1974–1979 (New Delhi: Controller of Publications, 1976), p. 7.

TABLE 2.3

Summary Profile of Growth of Agricultural Development at the District Level, from 1962/63–1964/65 to 1970/71–1972/73

Annual/Compound Growth Rate of Gross Value of Output (percent)	Cumulative Percentages of Total					
	Gross Cropped Area	Aggregate Output	Consumption of Nitrogen, Phosphorous, and Potassium	Use of Tractors	Pump Sets Installed	Gross Irrigated Area	Number of Districts in India

	Gross Cropped Area	Aggregate Output	Consumption of Nitrogen, Phosphorous, and Potassium	Use of Tractors	Pump Sets Installed	Gross Irrigated Area	Number of Districts in India
11.00–11.35	0.62	0.15	0.02	0.84	0.08	0.09	0.36
9.00–10.99	1.38	0.98	1.22	2.89	1.26	1.19	1.42
7.00–8.99	7.93	9.97	14.13	32.47	12.47	16.28	6.38
5.00–6.99	13.89	17.03	20.81	46.46	20.13	24.37	12.41
3.00–4.99	29.60	36.13	38.99	67.72	34.68	45.53	29.08
1.00–2.99	60.58	67.75	66.24	83.74	66.63	71.90	62.41
0.00–0.99	73.09	80.98	81.92	90.74	80.69	83.81	75.18

Note: Growth rate has been computed by valuing output in 1962/63 to 1964/65 and 1970/71 to 1972/73, at average all-India prices for each crop for the triennium 1970/71 to 1972/73.

Source: Centre for Regional Development, Jawaharlal Nehru University—Perspective Planning Division, Planning Commission, Project on Levels of Agricultural Development in India, reproduced from Government of India, Planning Commission, Fifth Five Year Plan, 1974–79 (New Delhi: Controller of Publications, 1976), p. 7.

The figures in Table 2.3 show that 12 percent of the districts in India, covering 14 percent of the total cropped area, accounted for approximately 20 percent of the total consumption of major inputs. Policies for agricultural development would have to be based on the perspective exhibited by figures in these tables. It is now generally accepted that growth in agricultural output can take place only through a rational and planned exploitation of ground and surface water, extension of new technologies in agriculture, and better distribution of inputs, such as fertilizers, at reasonable prices.

The chapter on "Perspective" for the Fifth Five-Year Plan estimated the growth rate in total crop area in the period 1961/62-1972/73 to be 0.54 percent compounded annually.[1] Using the elasticity of multiple cropping with respect to gross irrigated area, the National Commission on Agriculture estimated a growth rate of 0.66 percent compounded annually in the total cropped area from 1970/71 to the year 2000. At the same time, it has been estimated that the elasticity of total cropped area with respect to gross irrigated area for the entire nation is 0.20. The fifth plan assumes an increase of 4 percent per annum in the total irrigated area of the country, and the total cropped area is expected to expand by 0.7 percent per annum during the Fifth Five-Year Plan.

Geological surveys are being intensified and extended to different parts of the country to arrive at estimates of ground water resources. So far, approximately 65 percent of the recoverable area has not been investigated. The Fifth Five-Year Plan assumed an ultimate potential of groundwater at a conservative figure of 35 million hectares. The fifth plan has also stepped up considerably the allocation of funds for systematic evaluation of the country's groundwater resources. Table 2.4 gives the position of systematic hydrogeological survey work done in the country, indicating for each state the gap that exists that has to be made good by efforts in subsequent years.

Agricultural development in the country is essential for economic, political, and social stability. With higher economic growth, the demand for agricultural products will naturally increase at rates greater than the corresponding increase of population. It is estimated that in 1983/84, the total "requirement" of food grains will be approximately 143.5 million metric tons, and by 1988/89, it will be 161 million metric tons. This assumes per capita consumption levels of 180 and 190 kilograms in 1983/84 and 1988/89, respectively.

Issues pertaining to energy consumption and economic development are involved in the increase of nonfood grain crop production also. Sugarcane and cotton production are likely to benefit from improved irrigation facilities as well. Estimates for the fifth plan period indicate a growth rate of 3.94 percent per annum for nonfood grain crops, rising to 4.96 percent per annum during the seventh plan peri-

TABLE 2.4

Systematic Hydrogeological Survey, as of January 1, 1975
(in square kilometers)

Region/State	(1) Coverable Area	(2) Survey Completed	Gap (1) - (2)	Percent
Northern region				
Uttar Pradesh	271,293	170,070	101,223	37.3
Northwestern region				
Jammu and Kashmir	24,926	10,550	14,376	57.6
Delhi	1,485	1,483	2	0.1
Punjab	50,362	41,715	8,647	17.2
Haryana	44,222	40,190	4,032	9.2
Chandigarh	115	115	—	0.0
Himachal Pradesh	19,453	3,900	15,553	79.9
Western region				
Rajasthan	342,214	239,515	102,699	30.0
Gujarat	195,984	69,175	126,809	64.7
Eastern region				
Bihar	173,876	43,870	130,006	74.7
West Bengal	87,743	72,140	15,603	17.8
Orissa	155,782	34,845	117,939	75.7
Andaman and Nicobar	8,293	2,200	6,093	73.4
Northeastern region				
Assam	78,523	21,820	56,703	72.2
Meghalaya	22,489	50	22,439	99.7
Arunachal Pradesh	48,738	20	48,718	99.9
Tripura	10,477	2,550	7,927	75.6
Nagaland	14,367	600	13,767	95.8
Mizoram	21,496	625	20,871	97.0
Manipur	21,087	—	21,087	100.0
Central region				
Madhya Pradesh	442,841	78,730	364,111	82.2
Maharashtra	307,762	60,240	247,522	80.4
Goa, Diu, and Daman	3,813	2,275	1,538	40.3
Southern region				
Andhra Pradesh	276,811	96,720	180,094	65.0
Tamil Nadu	128,769	45,975	82,974	64.2
Pondicherry	480	—	480	100.0
Kerala	38,759	20,080	18,679	48.1
Laccadive Islands	32	—	32	100.0
Karanataka	191,773	38,720	153,053	79.8
Total	2,983,968	1,101,173	1,882,795	63.0

Source: Government of India, Planning Commission, Fifth Five Year Plan, 1974-79 (New Delhi: Controller of Publications, 1976), Annexure 1, p. 95.

od. With attendant growths in forestry, fishing, and animal husbandry, the annual rate of growth of the agricultural sector as a whole is expected to be 3.94 percent during the fifth plan and approximately 4.30 percent in the sixth and seventh plan periods.

The demand for fertilizers, being a demand derived from the demand for agricultural products, is tied indirectly with the increases in newer farming technologies. In terms of nutrients, demand for fertilizers is expected to increase to 4.80 million metric tons in 1978/79 and approximately 8 million metric tons in 1983/84. With the lead times required to establish production capacity, investments would have to be made commencing immediately to fulfill these levels of demand in the future. An area that is not often considered in projecting demand for the future is the price that farmers have to pay for fertilizers and the price of food grains that they produce. Recent years have shown a decline in the growth in fertilizer consumption as a result of a decreased margin between these two sets of prices.

The National Commission on Agriculture established forecasts for demand for industrial wood in the country. Forests as a whole occupy about 23 percent of the area of the country and contribute 1.4 percent of the net domestic product. There are considerable imbalances in the growth and exploitation of forest resources. These are brought about by a lack of transport, communications, and organizations to administer forestry programs effectively.

The three leading sectors that contribute commercial energy in this country, namely coal, electricity, and petroleum, accounted for 3.96 percent of the gross value added in the nonagricultural sector in 1973/74. If the targets in the Fifth Five-Year Plan are realized, this share is likely to go up to 5 percent by 1978/79 and to 5.56 percent by the end of the sixth plan. The electrical energy sector will be dealt with in a separate chapter of this book, but it may be mentioned here that it is expected that a growth rate of 8.5 to 9.5 percent per annum will probably be maintained over the next decade or so according to the Planning Commission. The consumption of refinery products grew at an average rate of 8.5 percent per annum compounded in the period 1960-73. The consumption in 1974/75 was nevertheless reduced to 1972 levels. With current indications, it appears that even the anticipated demand of 28.5 million metric tons in 1978/79 will not be realized, although substantial growth is likely to take place in sectors like fertilizers, transportation, irrigation, industry, and domestic energy consumption. Concern is often expressed in India, as it is all over the world, that our present rates of growth and consumption patterns will lead to rapid depletion of critical nonrenewable resources. It is often felt that technologies being adopted by India in its efforts toward economic growth are detrimental to its long-term interests. The elasticity of mineral consumption with respect to growth of GDP or manufacturing activity in India has been found to be greater than unity. This, however, is no greater than that which was found in other countries that went through similar stages of industrialization. Table 2.5 shows the long-term perspective of the availability of major minerals in the country based on estimates of depletion rates in 1988/89. The figures given show that for a number of critical minerals, such as chromite, kyanite, barite, and manganese, the country will have depleted its known resources before the year 2000. The position for copper and zinc does not appear to be much brighter; perhaps a long-term optimal policy for exploitation will have to be evolved in the years to come.

As mentioned before, the critical factor in India's development efforts is the growth rate of its population. The national population policy enunciated by the new government of India lays down a target for birthrate of 30 per 1,000 by March 1979 and 25 per 1,000 by March 1984. The Planning Commission has laid down a target growth rate of 1.14 percent by the end of the sixth plan period. To achieve this target, a number of measures will have to be adopted: revising the minimum age for marriage; greater education of females; spread of information on family planning; strengthening of research in reproductive biology and contraception; and providing incentives for individuals, groups, and communities. In terms of the national population policy, an estimate of 725.4 million by 1988/89 and 744.8 million by 1990/91 was obtained for India's future population. This included an estimated 545.1 million rural and 180.3 million urban population in 1988/89.

The Planning Commission has also worked out the projected mix of sectoral activities in the economy by the end of the Fifth Five-Year Plan. These are shown in Table 2.6, wherein projected rates of growth for the five-year plans are also included. Major increases in production are expected to take place in mining, chemical products, basic metals, electricity, and coal and petroleum products. The levels of increase projected for transport, other services, and agriculture are somewhat modest, which would lead to a decline in the share of agriculture in total output from 50.78 to 48.15 percent. The physical output targets laid down for major items of production in the Fifth Five-Year Plan are projected in Table 2.7.

As in other developing countries, a major constraint that inhibits economic development is the lack of financial resources. With the reduction in funds from external sources that has progressively taken place in the past few years, it is expected that in the Fifth Five-Year Plan period, 81.7 percent of the total resources required for financing the plan will be provided from domestic sources. External assistance is anticipated to be 14.9 percent of the total, and the balance of 3.4 percent is likely to be provided through deficit financing. These esti-

TABLE 2.5

Life of Known Reserves at 1988/89* Consumption Levels

Minerals	Balance in Years
Coking coal	44
Noncoking coal	
Indigenous	168
Plus exports	159
Iron ore-hematite	
Indigenous	165
Plus exports	62
Iron ore-magnetite	84
Manganese ore	
Indigenous	26
Plus exports	12
Chromite	
Indigenous	47
Plus exports	13
Bauxite	
Indigenous	66
Plus exports	45
Zinc	
Indigenous	—
Plus imports	11
Copper	
Indigenous	17
Plus imports	36
Lead	
Indigenous	29
Plus imports	—
Rock phosphate	
Indigenous	—
Plus imports	12
Limestone	475

*The fiscal year 1988/89 will be the final year of the Seventh Five-Year Plan.

Source: Government of India, Planning Commission, **Fifth Five Year Plan, 1974-79** (New Delhi: Controller of Publications, 1976), Table 4, p. 13.

TABLE 2.6

Projected Sectoral Rate of Growth in Gross Value of Output and Gross Value Added at Factor Cost for the Fifth Five-Year Plan and Sectoral Composition of Gross Value Added in 1973/74 and 1978/79

Sector	Percentage Average Annual Rate of Growth, 1978/79 over 1973/74		Composition of Gross Value Added at 1974/75 Prices	
	Value of Output	Value Added	1973/74	1978/79
Agriculture	3.94	3.34	50.78	48.15
Mining and manufacturing	7.10	6.54	15.78	17.49
Mining	12.58	11.44	0.99	1.37
Manufacturing	6.92	6.17	14.79	16.11
Food products	4.63	3.73	2.13	2.07
Textiles	3.45	3.21	3.50	3.31
Wood and paper products	6.75	4.90	0.58	0.59
Leather and rubber products	5.50	2.47	0.16	0.15
Chemical products	10.84	10.46	1.84	2.44
Coal and petroleum products	7.63	7.90	0.23	0.27
Nonmetallic mineral products	7.40	7.33	1.58	1.82
Basic metals	14.12	13.40	1.09	1.65
Metal products	5.60	4.64	1.08	1.09
Nonelectrical engineering products	8.40	7.99	0.61	0.73
Electrical engineering products	7.64	6.42	0.23	0.27
Transport equipment	3.73	3.12	0.96	0.90
Instruments	5.39	4.45	0.03	0.03
Miscellaneous industries	6.75	4.42	0.38	0.38
Electricity	10.12	8.15	0.79	0.94
Construction	5.90	5.18	4.06	4.21
Transport	4.79	4.70	3.43	3.48
Services	4.88	4.80	25.16	25.73
Total	—	4.37	100.00	100.00

Source: Government of India, Planning Commission, **Fifth Five Year Plan, 1974-79** (New Delhi: Controller of Publications, 1976), Table 1, p. 25.

TABLE 2.7

Actual Physical Output Levels, 1973/74, and Projections of
Physical Output Levels, 1978/79

Item	Unit	1973/74	1978/79
Food grains	mmt	104.70	125.00
Coal	mmt	79.00	124.00
Iron ore	mmt	35.70	56.00
Crude petroleum	mmt	7.20	14.18
Cotton cloth			
Mill sector	mm	4,083.00	4,800.00
Decentralized sector	mm	3,863.00	4,700.00
Paper and paperboard	tmt	776.00	1,050.00
Newsprint	tmt	48.70	80.00
Petroleum products (including lubricants)	mmt	19.70	27.00
Nitrogenous fertilizers (N)	tmt	1,058.00	2,900.00
Phosphatic fertilizers (P_2O_5)	tmt	319.00	770.00
Cement	mmt	14.67	20.80
Mild steel	mmt	4.89	8.80
Aluminum	tmt	147.90	310.00
Copper	tmt	12.70	37.00
Zinc	tmt	20.80	80.00
Electricity generation	GWH	72.00	116.00–117.00
Originating traffic in railways	mmt	—	260.00

Note: mmt = million metric tons; mm = million meters; tmt = thousand metric tons; GWH = billion kilowatt hours.

Source: Government of India, Planning Commission, Fifth Five Year Plan, 1974–79 (New Delhi: Controller of Publications, 1976), Table 2, p. 26.

mates, however, are likely to be changed with the new priorities that the recently installed government is likely to formulate.

An essential prerequisite for growth in developing economies is the ability of individuals and existing institutions to generate savings for investment. It is expected that during the fifth plan period, about 58 percent of the total investment will be in the public sector and the balance, 42 percent, in the private sector. The details of estimates of domestic savings by those sectors that generate these savings are given in Table 2.8. Of the total savings of Rs 583,200 million, about 27 percent, that is, Rs 159,940 million, will be contributed by the public sector, including government administrations, departmental and nondepartmental undertakings, and public financial institutions. The balance (about 73 percent) is accounted for by the private sector, including corporate enterprises, cooperatives, and households. The Fifth Five-Year Plan has been formulated on the basis of an estimated rise in savings from 14.4 percent of the GNP in 1973/74 to 15.9 percent in 1978/79. The breakup of savings against the above-mentioned sectors, that is, government, autonomous public enterprises, corporate enterprises, cooperatives, and households, is shown in Table 2.9.

An important element in India's plans for economic development is the foreign sector of its economy. Exports in the last five years have shown considerable buoyancy, rising to Rs 23,390 million in 1974/75 and to Rs 39,420 million in 1975/76. As against this, imports were Rs 45,190 million in 1974/75 and Rs 29,550 million in 1973/74. Imports rose further to Rs 51,580 million in 1975/76, showing an increase of 14 percent over the previous year. This was due mainly to an increase in industrial activity and a spillover of imports of food grains (a decision had been taken about this during the preceding years

TABLE 2.8

Domestic Savings by Generating Sectors

Sector	Savings (rupees in crores)*
Public sector	15,028
Central and state budgets	8,536
Central and state nondepartmental enterprises	6,492
Financial institutions	1,263
Reserve Bank of India	841
Other	422
Private sector	42,029
Private corporate nonfinancial sector	5,373
Cooperative noncredit institutions	175
Household sector	36,481
Total domestic savings	58,320

*One crore equals 10 million.

Source: Government of India, Planning Commission, Fifth Five Year Plan, 1974-79 (New Delhi: Controller of Publications, 1976), p. 41.

TABLE 2.9

Domestic Savings, by Sector of Origin, 1973/74 and 1978/79

	Savings (rupees in crores)*		Percent of GNP	
	1973/74 (at 1973/74 prices)	1978/79 (at 1975/76 prices)		
Sector			1973/74	1978/79
Public sector	1,423	4,045	2.5	4.6
Government	772	2,704	1.4	3.1
Autonomous public enterprises	651	1,341	1.1	1.5
Private sector	6,824	9,868	11.9	11.3
Corporate	821	1,268	1.4	1.4
Cooperative	65	95	0.1	0.1
Household	5,938	8,505	10.4	9.8
Total	8,247	13,913	14.4	15.9

*One crore equals 10 million.

Source: Government of India, Planning Commission, Fifth Five Year Plan, 1974-79 (New Delhi: Controller of Publications, 1976), p. 42.

of drought and scarcity). The balance of trade for the country has deteriorated since 1973/74, mainly because of large-scale increases in the prices of food, fertilizers, and petroleum. The unit value indexes of imports of these three items went up from 182, 91, and 334, respectively, in 1973/74 to 229, 173, and 736, respectively, in 1974/75 and to 276, 167, and 829, respectively, in 1975/76. The terms of trade index compiled by the government of India fell sharply from 124 in 1972/73 to values of 106, 77, and 70 in 1973/74, 1974/75, and 1975-76, respectively. Fortunately, the country was able to achieve a large increase in foreign exchange reserves during recent years, mainly due to receiving higher remittances from private sources. In the year 1975/76, total reserves went up to Rs 18,850 million, which showed an increase of Rs 9,160 million over the reserves in the previous year.

Restoring health to the foreign sector of the economy will need major efforts on the part of the country to increase exports and reduce imports. The target for an increase in exports has been laid down as 8.5 percent per annum for the plan period. The largest group of items in the export of the country by the end of the fifth plan are expected to be engineering goods. The optimism about exports of this group of items is based mainly on the extensive opportunities for sale of engineering goods in surrounding countries and the diversification of production facilities and capabilities within the country.

The plans for investments in the fifth plan period are designed to lead toward import substitution in four major areas of economic activity. These are energy, production of metals, fertilizers, and agriculture. In the case of energy, there will be efforts at increased oil exploration, a movement toward greater consumption of coal, and increased exploitation of hydroelectric potential. In the field of metals, steel imports will be minimized and restricted to some very special categories, and in the case of fertilizers, an expansion of capacity and production levels is likely to bring down imports in the next few years. Projections of exports and imports are shown in Tables 2.10 and 2.11, respectively.

Industry in India has reached a stage where, given growth-oriented policies, a satisfactory expansion in output will continue to take place in the years to come. Some of the directions in which industrial growth is being pushed through centrally planned efforts can be assessed by looking at the outlays that are expected to be made during the period of the Fifth Five-Year Plan period. These are shown in Table 2.12.

It is expected that the demand for steel by 1978/79 will be approximately 7.75 million metric tons, whereas production is likely to be 8.8 million metric tons. Current indications, however, point to a lower likely offtake in 1978/79, thereby making it possible that the surplus for export will be larger than indicated by the above figures. The largest steel plant in the country (at Bokaro) is expected to expand its production to 4 million metric tons by the middle of 1979. Work has also started on the expansion of the Bhilai steel plant, which will increase its production to 4 million metric tons by the end of 1981. Currently, a major debate is taking place on the long-term perspectives of the steel industry. The former minister for Steel in the previous government had propagated the idea of planning for a total capacity of 75 million metric tons by the turn of the century. This figure is now considered unrealistic and is likely to be revised downward. Preliminary work and preinvestment planning must start early, in view of the long gestation period of steel plants, if steel-making capacity is to grow substantially in the 1980s.

In nonferrous metals, the public sector plant at Korba is likely to reach its full capacity of 100,000 metric tons by the end of the fifth plan. This would provide a total capacity for aluminum in the country of 325,000 metric tons, the balance being provided by private sector plants in the country. For copper production, the fifth plan

TABLE 2.10

Export Projections for the Fifth Five-Year Plan Period
(rupees in crores)[a]

Item	Draft Plan	Revised Projections
Tea	840	1,233
Jute products	1,200	1,317
Coffee	190	368
Tobacco products	335	550
Oil cakes	315	481
Cashew kernels	405	632
Spices	170	365
Raw cotton	115	75
Fish and fish preparations	580	853
Sugar	115	1,424
Iron ore	980	1,373
Coal	40	75
Mica and mica products	120	220
Cotton textiles—mill-made[b]	1,000	1,585
Handloomed piece goods	155	256
Coir yarn and goods	90	131
Fabrics of man-made fibers	80	143
Leather and leather goods, including footwear	945	1,352
Chemical and allied products	370	567
Rubber	60	88
Engineering goods	1,500	2,328
Iron and steel	240	786
Handicrafts	905	1,237
Pearls, precious stones	600	695
Other handicrafts	305	542
Total	10,750	17,439
Other	1,830	4,283
Grand total	12,580	21,722

[a]One crore equals 10 million.
[b]Including piece goods, yarn, apparel, hosiery, and other cotton manufactures.

Source: Government of India, Planning Commission, <u>Fifth Five Year Plan, 1974-79</u> (New Delhi: Controller of Publications, 1976), Annexure 15, p. 113.

TABLE 2.11

Import Projections for the Fifth Five-Year Plan Period
(rupees in crores)[a]

Item	Draft Plan	Revised Projections
Metal, ores, and scrap	1,920	2,347
Metal products, machinery, and transport equipment, including components and spare parts	4,010	6,034
Petroleum crude, products, and lubricants	3,080	6,280
Fertilizers and raw materials for fertilizers	1,450	3,168
Other[b]	3,640	10,705
Total	14,100	28,524

[a]One crore equals 10 million.
[b]Revised estimates include provisions for food grains imports.

Source: Government of India, Planning Commission, Fifth Five Year Plan, 1974-79 (New Delhi: Controller of Publications, 1976), Annexure 16, p. 114.

will probably establish a total capacity of 37,000 metric tons of copper from domestic ore by the end of the plan period. Similarly, capacity for zinc production is likely to increase to 95,000 metric tons by 1978/79.

For the engineering goods industry, the major chunk of investment in the fifth plan will be made for stepping up output of electric generation equipment and various types of machine tools. This will be done largely in the public sector units of Bharat Heavy Electricals Limited (BHEL) and Hindustan Machine Tools. The plan also aims at stepping up production of scooters (under the public sector) in a big way, with the concept of a large mother unit supplying components for assembly to a number of subsidiary units in different states. A major expansion program is also in hand for the Hindustan Shipyard to achieve production of three ships per annum of 21,600 deadweight tons and the Cochin Shipyard for two ships per annum of 75,000 deadweight tons by 1977/78. In the final year of the plan, it is expected that output will be increased.

The fifth plan aims to achieve a total production of 56 million metric tons of iron ore. The largest project being executed in this

TABLE 2.12

Outlays for Important Industries, 1974/75–1978/79
(rupees in crores)*

Industry	Outlay
Steel	1,675
Fertilizers	1,533
Coal (including lignite)	1,147
Oil exploration, refining, and distribution	1,575
Petrochemicals	349
Machinery and engineering industries	365
Nonferrous metals	468
Iron ore (including Kudremukh project)	513
Paper and newsprint	203
Cement	102
Textiles	104
Shipbuilding	147

*One crore equals 10 million.

Source: Government of India, Planning Commission, Fifth Five Year Plan, 1974-79 (New Delhi: Controller of Publications, 1976), p. 61.

period will exploit the magnetite deposits in the Kudremukh area in South India for a production of 7.5 million metric tons of magnetite concentrates at a total investment of approximately Rs 6 billion. This project is being jointly financed with Iran, with guaranteed marketing arrangements for sale of the product to Iran.

In 1973/74, India produced 7,900 million meters of cloth. This is likely to go up to 9,500 million meters in 1978/79; of this, 4,800 million meters will be in textile mills, with the balance in the decentralized sector. Again, the present government is likely to provide greater encouragement to production of cloth in the decentralized sector, but firm plans have not yet materialized. The capacity for producing cement in the country is likely to increase from 17.2 million metric tons in 1973/74 to 23.5 million metric tons in the year 1978/79. Another energy-intensive process that is likely to increase substantially in the fifth plan is the production of paper and paperboard. A total provision of Rs 203 crores has been made for the development of the paper and newsprint industry. Major expansion will also take place in the form of completion of heavy water plants, various schemes

INDIAN ECONOMY   41

under the Nuclear Fuel Complex, and other public sector undertakings under the Department of Atomic Energy (DAE). A total allocation of Rs 184.18 crores has been made for these schemes.

Major expansion schemes are being implemented in the transport sector as well. These include investments in the Indian Railways, as well as in the roadway network in the country. Details of these are given in Tables 2.13 and 2.14. By 1978/79, the railways will be equipped to carry an estimated originating freight traffic of 250 to 260 million metric tons. Coal will represent the largest commodity, totaling approximately 98 million metric tons of originating traffic. With the Indian Railways' production program, a large number of locomotives, freight cars, and passenger cars will be replaced during the period. Schemes are also in hand to provide metropolitan railway transportation in the major cities of the country. The largest of these on which work is proceeding is in and around the city of Calcutta.

In the scheme for development of road transportation, the major investment will be made to complete schemes that were undertaken in the Fourth Five-Year Plan. In Table 2.14, the figures in brackets represent spillover schemes that were already in hand at the start of the Fifth Five-Year Plan.

The figures mentioned above, pertaining to some key sectors in the Indian economy, are presented for the purpose of showing the scale of growth that is contemplated in the next two years and for arriving at some tentative implications that these growth rates will have on the energy situation in the country. A detailed survey of the past five-year plans and the Fifth Five-Year Plan has been presented mainly to provide a backdrop for energy policies, which must essentially dovetail with developmental efforts in all other sectors.

A popular misconception in the West is that India has a poor industrial base. It must be emphasized that even though in terms of share of the total GNP. agriculture is still the major economic activity within the nation, the growth of industry in recent years has been quite spectacular, taking India to the position of a major industrial nation. Even though the Fifth Five-Year Plan extends only up to the year 1978/79, the investments and efforts being made in this period will set a stage for growth in subsequent years. Policies formulated and implemented by the government affect the living styles and patterns of energy consumption in the country at large. It would, therefore, be useful to explore some of the social aspects of Indian life, particularly in the rural sector of the country, before concluding our analysis of economic development in India.

SOCIOECONOMIC DEVELOPMENT

Attempts at modernization in India have often yielded little or no success. Among the various constraints that inhibit the technologi-

TABLE 2.13

Revised Fifth Five-Year Plan Outlay on Railways
(rupees in crores)*

Program	Expenditure (1974–77)	(1977–79)	(1974–79)
Rolling stock	556.8	500.0	1,056.8
Workshops/sheds	35.9	42.0	77.9
Machinery and plant	23.7	18.0	41.7
Track renewals	104.1	105.0	209.1
Bridge works	23.3	24.0	47.3
Line capacity works	169.9	146.0	315.9
Signaling and safety	39.2	32.0	71.2
Electrification	59.1	42.0	101.1
Other electrical works	13.0	10.0	23.0
New lines	55.2	42.0	97.2
Staff quarters	15.2		
Staff welfare	8.6	31.0	67.2
Users' amenities	7.0		
Other specified works	5.4		
Investment in road services	22.7	26.0	48.7
Inventories	−15.3	10.0	−5.3
Total	1,123.8	1,028.0	2,151.8
Metropolitan transport	25.2	25.0	50.2
Grand total	1,149.0	1,053.0	2,202.0

*One crore equals 10 million.

Note: At the time of finalization of the plan, in 1976, expenditures had already been incurred or authorized as indicated by figures in the first column. The balance that has been allocated for the remainder of the plan period is indicated in the second column.

Source: Government of India, Planning Commission, Fifth Five Year Plan, 1974–79 (New Delhi: Controller of Publications, 1976), Annexure 35, p. 152.

cal change—or, in fact, change of any sort—often the biggest are social and sociological in nature. Various indexes can be considered for measuring modernization. Based on the experience of other countries, one could perhaps postulate that urbanization, the development of transport and communications, reduction of disparities in income, spread of education and medical care, and so forth, are essential yardsticks employed in assessing overall social progress. Often, purely economic indexes, such as per capita income, per capita consumption of food grains, electricity, and so forth, are considered in isolation of some of the factors mentioned above. Patterns of growth in the initial stages of development in countries such as India are often accompanied by large variations in the rates of growth exhibited by any selected set of indicators. This happens quite often because of a pronounced duality in society. Sociologists have sometimes believed that Western education and values can spark social change in developing countries. The experience of India and a host of other

TABLE 2.14

Revised Outlays for Central Programs

Program	Estimated Expenditure (1974/77)	Outlays (1977/79)	Revised Plan Outlay
National highways	176.56	151.06	327.62
	(159.79)*	(93.06)	(252.85)
Strategic roads	14.00	24.00	38.00
	(12.00)	(21.00)	(33.00)
Roads of interstate and economic importance	9.24	20.76	30.00
	(9.24)	(14.76)	(24.00)
Highway research and development	0.20	1.80	2.00
Roads in sensitive border areas	1.00	9.00	10.00
Special road/bridge works of national significance, mainly for second Hooghly Bridge at Calcutta	9.02	15.98	25.00
Tools and plants	7.82	5.00	12.82
Total	217.84	227.60	445.44
	(181.03)	(128.82)	(309.85)

*Figures in parentheses refer to spillover schemes.

Note: At the time of finalization of the plan, in 1976, expenditure had already been incurred or authorized as indicated by figures in the first column. The balance that has been allocated for the remainder of the plan period is indicated in the second column.

Source: Government of India, Planning Commission, Fifth Five Year Plan, 1974-79 (New Delhi: Controller of Publications, 1976), p. 70.

countries, however, has not shown this to be the case. In a study carried out by David McClelland,[2] he found that children going to schools where the classes were taught in the vernacular had a higher achievement motivation than those attending schools where English was the medium of education (these were run by Christian missionaries). The students in the latter schools often came from parents in high-income brackets, who followed Western styles of living and thinking. Other scholars have also found that high motivational levels are not exclusively the preserve of the socially elite. The inherent fatalistic and passive concepts attributed to Indian society could very well be the result of lack of opportunities for advancement rather than vice versa. Given appropriate opportunities, motivational levels have been seen to rise in response to possibilities for advancement.

Modernization and the spreading of modern ideas in any country generally proceed from demonstration and practice by a nucleus of dynamic individuals in that particular society. This often happened in Western societies through a growth of the middle class. The small percentage of the middle class in developing societies, however, leaves a gap, which is often filled by society looking toward the West, particularly for leads in technological development.

The extremely poor class in rural society had no choice but to move toward urban centers for employment in Western societies. Due to the lack of infrastructure and poor resources in developing societies, it is generally the relatively affluent individuals in rural areas that have opportunities for movement and assimilation in urban areas. As such, with industrial growth, the process of disparate growth gets accentuated, and the alienation of the vast rural mass from the urban elite is accelerated.

The habits of the colonial past are also difficult to eliminate, particularly for a country that had to import most of the commodities consumed by it in the past. Efforts at import substitution, as a part of development, therefore, naturally evolved as a replacement for imports that had been taking place hitherto. This was another factor that resulted in continued technological dependence on the West in the 30 years since India's independence. Imbalances in income distribution necessarily continued, and in fact were accentuated, after independence, since the benefits of growth generally went to a small section of society.

One of the major problems faced by India in its developmental efforts (and in fact in just trying to hold together as a nation) has been the existence of a variety of communities in the country. Communities that had access to the colonial-based economy were the ones that tasted relative affluence in the early stages of development. As an example, the State of Bengal, by virtue of the British presence in Calcutta, was the first to take advantage of the effects of trade, administration, and education. The result was that for a number of years, Bengalis were foremost in occupying important administrative posts in the government, as well as in the private sector, which was then in the hands of the British. Again, Hindus in Bengal very soon went far ahead of the Muslim masses, who were engaged largely in agricultural tasks (a large percentage had been converted from low-caste Hindu communities in earlier times). The Brahmins of West Bengal, with a tradition of learning and culture, found it fairly simple to change over to English and adapt themselves to the needs of the colonial powers, whereas the lower castes, bound by feudal traditions, remained below the poverty line. The Hindu community very soon seized advantage and took control of most of the trading activities. Since power in the country had been taken by the British from the hands of Muslim rulers, very soon, Muslim centers of power and trade declined at the cost of new colonial towns, which were flourishing. The 1931 census illustrates how the important centers of commerce in that period became strongholds of Hindu populations, whereas the old centers of commerce and industry retained a greater Muslim population. This was in spite of the fact that the total Hindu population of the country was almost five times that of the Muslims.

Communal disparities took their toll in the form of the partition of the country into two nations, namely, India and Pakistan. This division left permanent scars and a deep impact on the postindependence history of the two nations. A mingling of the two cultures and a breakdown of communal barriers that might have taken place did not, since communal distinctions were not only preserved but strengthened by the British attitude of "divide and rule." Sociological forces and political compulsions that might have brought about assimilation of different communities in the country did not take place. Attitudes and traditions have remained hardened, and the desired transformation into a modernized culture has not taken place as rapidly as it might have under different circumstances.

Even though independence in a number of nations has brought about rapid social change, the (status quo) attitudes that were enshrined in the communal and social rigidities of the Indian population have not allowed for a suitable transformation. Tradition has, therefore, reigned supreme. Economic activities and life-styles that existed for centuries continue to do so in the rural areas of the country. Government policies for economic development and related energy policy have to contend with the constraints and inhibitions of social and cultural tradition.

Consumption of rural energy and its pattern of usage are similarly resistant to change. A study carried out by the National Council of Applied Economic Research (NCAER) published in 1965 projected patterns of energy consumption for the country for the year 1975/76 based on detailed surveys of a number of villages.[3] It is interesting to observe that even though the levels of consumption projected in this study have not been realized, the pattern and mix that the council projected has not changed materially. This indicates that while income changes did not take place to bring about greater consumption of energy, relative prices and availability of alternative energy in the rural sector did not change either to result in a change in consumption patterns. It is also interesting to notice that among their recommendations, the usual concerns about protecting forest resources of the country, installing a large number of gobar gas plants,* using vegetable waste, extending rural electrification, and marketing coal in rural areas were discussed. Programs on these lines have been launched at different levels, but they have obviously had only a minimal impact on the rural energy scene. It is also a sad commentary on the priorities laid down by policy makers that few comprehensive and in-depth studies on rural energy demand and supply have been made in recent years, except for the work of the FPC. Rural India continues to consume almost 50 percent of the energy used in the country, but it continues to suffer from the neglect to which scholars, academicians, and policy makers have assigned rural problems.

NOTES

1. Government of India, Planning Commission, Fifth Five Year Plan, 1974-79 (New Delhi: Controller of Publications, 1976), p. 5.
2. David McClelland, The Achieving Society (New York: D. Van Nostrand, 1961), p. 414.
3. National Council of Applied Economic Research, Domestic Fuels in Rural India (New Delhi: National Council of Applied Economic Research, 1965), pp. 42-62.

*Gobar gas is cow dung gas. The plants are more formally referred to as biogas plants.

CHAPTER 3
ENERGY DEMAND

National or regional energy policy is established on the basis of estimates of current and future demands for various forms of energy. The whole area of forecasting of energy demand has not reached maturity, since this is a relatively new field in which sophisticated forecasting skills have been applied only in recent years. One reason why detailed forecasts based on sound quantitative techniques have not been developed and compiled in the past is because relative prices of most commercial forms of energy have remained reasonably stable over the years. In addition, the real price of various sources of energy declined at fairly stable and predictable rates, thereby making trend forecasts and simple extrapolation reasonably reliable for planning purposes. All this has changed with the rapid increases in prices of crude oil all over the world. The price increases in petroleum have necessarily had an impact on other energy sources as well. For instance, wherever fuel oil is used for power generation, the cost of power generated has gone up correspondingly. In other cases, also, such as where coal is used as a fuel or is converted into power generated from steam-electric sources, the increased demand for other energy forms as a substitute for petroleum products is leading to larger demand for substitute forms. This has inevitably led to increasing costs of production, since supply curves for energy, as in the case of other commodities, are upward sloping.

In some respects, the problem of forecasting demand for energy differs considerably for developing countries from that in developed industrialized societies. This arises not only because of data limitations but also because of the scope of government policy in influencing demand and consumption. In India, for instance, the Fifth Five-Year Plan (which is currently in effect) envisages electrification of only about 30 percent of villages in the country. The rate of rural electrification is determined by political factors and government decision making. To the extent this program is accelerated, the demand for electricity for rural consumption will also increase.

Pricing of petroleum products is again an area in which government policy has a significant effect, since production and distribution of petroleum and petroleum products is mainly in the public sector. Prices are generally set by arriving at taxation levels that will generate desired revenues for various developmental activities and meet fiscal budgetary requirements, as well as keep in view the effects of prices on total consumption of petroleum products. In certain cases, particularly when shortages of various energy forms develop, the government resorts to various measures involving mandatory allocations. In the period 1972/73 to 1974/75, various parts of the country suffered from different degrees of shortages of power. In such circumstances, central, state, and local governments generally lay down limits for power supply and effectively control the amount of power or energy consumed by different types of consumers.

Demand forecasting in the energy sector is further complicated by the difficulties in predicting the movement of key economic-demographic variables, such as the rate of growth of population, the rural-urban mix of the total population, and increases in GNP and per capita income. These factors have a pronounced effect on the demand for energy.

Faced with the difficulties enumerated above, scholars and government committees have dealt with the problem of energy forecasting from various points of view and have used a variety of methodologies. For aggregate national policy making, a very broad perspective has, therefore, to be used to make any exercise in future planning relevant and meaningful. Any set of forecasts that is produced must be seen in conjunction with the supply constraints prevailing at the time, contingent on the various influencing factors taking on unanticipated values. A unique set of quantifiable forecasts would be difficult to accept as reliable and useful for planning in the energy sector. The FPC, which was engaged in the exhaustive task of investigating the energy sector of India and suggesting various policy measures, was overtaken by events that occurred in the years 1973/74 and 1974/75. In the absence of definite estimates of elasticities related to price, income, and interfuel substitution, their forecasts had to be modified, using some broad assumptions about the extent to which increased oil prices would affect demand in the future. The extent of substitution assumed by them was based on technological rather than economic possibilities. For instance, they determined the reduction in quantity demanded of fuel oil, if the use of coal in place of fuel oil was to be substituted in all furnaces and boilers in the industrial sector. Using similar assumptions for substitution, they worked out

ENERGY DEMAND

forecasts of demand for three scenarios, classified as Cases 1, 2, and 3. In the judgment of the FPC, Case 2 was regarded as the normal case, resting on a feasible level of substitution that could be achieved by government as well as private decision making. Before taking up some tentative estimates of future demand that have been worked out in various studies, including the author's, it would be useful to explore the determinants of demand for energy in different sectors.

DETERMINANTS OF DEMAND FOR VARIOUS ENERGY SOURCES

Kerosene oil among the commercial fuels has been used on a large scale for lighting and cooking purposes in the domestic sector. This, of course, has been decreasing in relative terms, as a result of greater consumption of electricity and other energy forms, both in the urban and rural parts of the country. However, rapid rates of urbanization often bring about an increase in the demand for kerosene oil, because rural consumers, who generally use noncommercial sources of energy, such as cow dung, wastes, firewood, and so forth, move to urban areas, where kerosene oil is used extensively for domestic purposes. The popularity of kerosene perhaps lies in the excellent distribution network that exists throughout the country and in the ease with which it can be stored and consumed in the household. Soft coke is a substitute for cooking purposes; it is not usually preferred because of its messy characteristics and the space it takes up in storage and in use in the household. The evolution and production of a smokeless variety of coke, which is being pioneered and marketed by Coal India, the major public sector undertaking dealing with coal, may show some results toward substitution in favor of coal in the urban areas in the next few years.

Unlike developed countries, where cooking and lighting are based on gas and electricity, a very small section of the Indian population in the larger urban areas of India use liquid petroleum gas (LPG) or electricity for domestic purposes. An increase in the number of the urban middle class will, therefore, lead to greater demand for LPG and electricity in the years to come.

The industrial sector of the country consumes the major share of commercial energy. Interfuel substitution in this sector could lead to changes in the pattern of demand. For instance, fuel oil can be substituted economically for coal wherever it is used for producing heat, except in a very small number of industries. Since production technologies employed in developing countries are generally based on those that have evolved in the industrialized world, patterns of energy consumption between industries in the developed and developing countries are quite similar. The growth of process industries accentuates this trend to a large degree, since these are generally capital- and energy-intensive by nature. It is unlikely that any major change in energy intensiveness in specific industries will take place in the near future. At best, mandatory controls and pricing policies may lead to more efficient use of energy rather than their wholesale substitution by labor or capital in the short and medium run.

The transport sector in India is another large consumer of energy. For road transport services based on private cars, taxis, and three-wheeler and two-wheeler vehicles, gasoline is consumed as the standard fuel. HSDO is consumed generally by commercial road vehicles and by locomotives. The growth in consumption of gasoline has slowed down substantially in the last three or four years. In fact, after increases in the excise tax on gasoline in 1974, there was an immediate reduction of 22 percent in overall consumption of gasoline in the country in the following year. Further, the light motor vehicles industry suffered a serious slump in the last three years due to the reduced demand for new vehicles, and as a result of this, the total road mileage in the next few years is likely to grow at a slower rate than during the past two decades. The Indian Railways is continuing a program of dieselization, which involves a gradual replacement of steam locomotives, which still form the bulk of the power on the Indian railway system with diesel locomotives. At the same time, electrification of major trunk routes is proceeding apace. With the increase of crude oil prices, the economies of scale that the railways offer in movement of freight is likely to result in a substitution from road transportation for long hauls in favor of rail transportation. This does not, however, mean that freight movement by road is likely to decline. The potential for short-haul traffic by road is very large. As rural areas are opened up by better roadway networks and a greater amount of commercial activity takes place, the roadways will have an increasing role to play in intraregional movements of commodities.

In recent decades, as mentioned before, the agricultural sector has contributed in increasing measure to the demand for energy. This has been in the form of HSDO and light diesel oil (LDO) for irrigation by pump sets, as well as for tractors and mechanized equipment, which are being used increasingly in farming activities. Even though with an increase of diesel oil prices there would normally have been a tendency for substitution from diesel-operated pumps to electric-operated pumps, difficulties in power supply experienced in recent years and the relatively slow pace of rural electrification have continued to result in a greater demand for diesel pump sets. In the year 1973/74, there was a total of approximately 1.5 million diesel pumps in operation, and it is expected that with the current rates of growth, these may increase by about 0.2 million pump sets every year.

ENERGY DEMAND

In considering the demand for various sources of energy, it would also be relevant to take into account the demand for various uses of petroleum and coal for nonenergy uses. This would be in the form of feedstock for fertilizer production, manufacture of petrochemicals, and so forth. The evolution of modern agricultural techniques and the large-scale benefits of petrochemical products is likely to lead to a much greater increase in the share of nonenergy uses in the overall demand for coal and oil in the future.

In Table 3.1, we have presented percentages for some of the different petroleum products currently consumed in the country as a share of the total consumption. The end use for each of these products is also indicated in the same table.

These percentages will change, of course, as newer products and their uses are discovered in the future. The traditional concept of demand in evidence in developed societies does not always operate in developing societies. Quite often, the demand for a particular product is not generated because of nonexistence or regional imbalances in supply. A large number of products produced by petrochemical industries are not known in rural areas because a sufficiently large production base has not yet been established (also due to the low purchasing power of the rural population); the distribution channels for these products, which may develop in the years to come, do not exist at present. The high growth rate recorded by the world petrochemical industry from a meager 50,000 metric tons of production in 1940 to about 50 million metric tons in 1970 was essentially due to the ease with which synthetic fibers and polymers can substitute and supplement natural fibers and other traditional materials. The average Indian still wears clothes made from pure cotton, and substitution by synthetic fibers has not been as rapid as in other countries, perhaps on account of the climatic conditions, which could make all except some very special types of synthetic fabrics uncomfortable to wear in most parts of India. The situation, of course, is changing rapidly. Political leaders in India who have generally favored the use of homespun cotton (a spillover from Gandhian times) are now openly advocating the use of synthetic fibers as an economically superior substitute. It is no longer unusual to find Indian farmers, particularly in the more affluent northern parts of the country, wearing polyester shirts.

A quarter century ago, 75 percent of the clothing needs of the world were met by cotton; man-made fibers accounted for only 14 percent and wool approximately 10 percent. In 1975, approximately 35 percent of the world's clothing needs were met by synthetics; this was due not only to a change in people's tastes but because clothing from synthetic fibers is a cheaper alternative to clothing produced from cotton. The following estimates provide some idea of the average extra resources required to produce each additional 1 billion meters of cloth through natural fibers (cotton) under Indian conditions:

Factors	Extra Resources
Investment (millions of U.S. dollars)	330
Extra spindles (millions)	2
Extra looms	33,000
Power (millions of kilowatts)	400
Fuel (metric tons)	33,000

TABLE 3.1

Consumption of Petroleum Products by End Use

Product	Type of Consumption	Energy or Nonenergy	Percentage of Total Consumption
Naphtha	Fertilizers and chemicals	Nonenergy	8
Gasoline	Automobiles	Energy	7
Jet fuel	Aircraft	Energy	4
Kerosene	Cooking and lighting in domestic sector	Energy	13
High speed diesel oil	Tractors, trucks, irrigation pumps, stationary power-generating equipment, and railways	Energy	22
Light diesel oil	Irrigation pumps, power-generating equipment, and fishing craft	Energy	6
Fuel oil	Power generation and boilers used in industry	Energy	26
Bitumen	Road building and construction	Nonenergy	5
Lubricating oil	Automotive and industrial lubrication	Nonenergy	9
Total			100

Source: Report of the Fuel Policy Committee (New Delhi: Controller of Publications, 1974), p. 14.

In contrast with this, the additional requirement for all viscose material would be 92 million kilograms of wood pulp. Of course, the most important effect of substitution from cotton to synthetics would be the release of large areas of land, which could be used for food grain production. Rayon production also would normally not be favored because it is based on food and forest resources and because rayon production is energy intensive. Synthetics from petrochemicals are, therefore, likely to grow at a very rapid rate in the future.

The use of plastics for various items of household consumption, as well as for agricultural and water management, also is likely to grow rapidly in the future. For instance, in a country to which water resources are vitally important, as in the case of India, thermoplastics can be used to prevent loss of water through seepage and evaporation. They can be used very effectively to convert arid zones and desert areas into fertile lands. One of the techniques used successfully in this effort is known as mulching, which involves the covering of irrigated fields by black polyethylene film sheets. This has a host of advantages, such as early maturing of crops, increase in average yield, better retention of soil moisture, reduced weed growth, reduced fertilizer consumption, and so forth. Fruits, vegetables, sugarcane, and corn are crops generally found suitable for mulching.

Another area where plastics have an important contribution to make is in the field of food grain storage and transportation. Large quantities of food grains in India are lost on account of moisture, insects, and rodents. Estimates on the extent of this loss vary from 5 to 25 percent of the total food grains produced. The use of thin plastic films between layers of clay or brick has been found very effective and economical in preventing losses of this type. With prices and costs at existing levels in India, a two-metric-ton brick-polyethylene storage bin would cost about U.S. $16, as compared to a steel bin of U.S. $133 and a cement concrete bin of U.S. $150. Another major use of polyethylene film in agricultural management is in the lining of canals. Low-density polyethylene (LDPE) used as canal lining can save between 30 to 60 percent water loss through seepage in some parts of the country. These films are easy to handle and much cheaper in the long run than concrete or brick lining. In some Indian canals, films laid as far back as in 1959 are still in good shape. Similarly, flexible tubes made out of LDPE or polyvinyl chloride (PVC), used in conveying water from a well head to the fields it serves, can conserve as much as 40 percent of water that would otherwise be lost in transit.

In irrigation, too, plastics have a vital role to play. Two systems of irrigation, known as "sprinkler irrigation" and "trickle irrigation," are being used successfully. In an experiment in the state of Rajasthan, the yield of wheat increased by about 35 percent where sprinkler irrigation was used instead of conventional border strip or check basin irrigation. In another trial, with trickle irrigation, the yield of certain vegetables increased by about 47 percent.

The possible developments are mentioned only to indicate that particularly in the field of petroleum and petroleum products, a major shift is likely to take place in the future in favor of greater production of petrochemical products that have nonenergy uses. Government policy, both through legislation and incentives, will undoubtedly lead to reduction in growth of petroleum and petroleum products for energy uses in the future. For instance, the use of fuel oils is likely to be eliminated in power generation and most industrial applications over the next decades. This observation emphasizes the effect of government policy in determining demand for energy in a country such as India, as opposed to the situation existing in free market economies in the developed countries of the West.

Studies in the past have often fallen into the trap of merely correlating levels of aggregate economic measures, such as GNP, with total energy consumption. Such an approach ignores the vital changes that are taking place in the pattern of energy consumption as brought about by changes in economic-demographic activities in the country. Very little attention has been systematically directed toward developments in the rural sector of the Indian economy and its impact on different forms of energy consumption. The problem of rural India is as much that of low consumption of energy as it is of inefficient consumption of energy. Domestic consumption of energy is related to the basic activities of cooking and heating in rural households. Data from India indicate that the fuel used for cooking in open hearths with slow burning fires is 5 to 7 million British thermal units (Btu) per capita per annum. This consumption level interestingly is higher than in the United States, where the annual per capita use for cooking purposes is 3 million Btu for electric stoves and 1 to 2 million Btu for gas stoves. This difference, of course, also arises because a great deal of food cooked in the United States is in partly or fully precooked condition and because the type of food that an Indian normally consumes requires slow and prolonged cooking. Some estimates indicate that the consumption of LP Gas for cooking in Indian cities is usually in the range of 1 million to 1.5 million Btu per capita. This indicates in relative terms the inefficient manner in which fuel is consumed in rural homes. The manner in which fuel is used for cooking in India is not only inefficient in terms of energy but, perhaps, in labor and capital as well. Apart from the large amount of labor that is used in the process of cooking, considerable labor is also expended in the collection of fuel wood and other types of fuel.

Very little fuel is used in rural areas for heating interior areas. Typically, a rural family in the mountainous regions of India would use a small stove with soft coke or wood for keeping a fire going in the household. In some places, such as in Kashmir, a small stove with a pot of embers is kept burning close to the body in the cold season. Again, this provides a very inefficient method of energy conversion and usage (not to speak of the health implications) both economically and thermally. The extreme poverty of some regions results in houses or huts being built with little or no insulation, and to compensate for this, the entire family collects firewood in large quantities to be burnt within the household.

Wherever commercial sources of energy have entered into rural areas, the problem of peak demand, particularly for electricity, makes the supply of such energy highly expensive. It is not unusual for two villages connected on a common distribution line for the supply of electric power to be served with staggered timings, so that there is no possibility of simultaneous demand and increased peak requirements. The peak demand for energy whenever it is related to agricultural activities results largely from the nature of farming undertaken in particular areas. Wherever single-cropping patterns are in practice, the demand for energy is concentrated over a very short period in the year. The problem of peak activities affects economic development in three ways, as identified by Arjun Makhijani,[1] which are mentioned below:

1. It results in high unemployment for a large part of the year, which is often the main reason for low productivity of labor.
2. It retards the introduction of innovations in agriculture, since the farmer does not have any possibilities for hiring additional labor, which may be required coexistently with peak demand periods.
3. It acts as a disincentive to reduction in birthrates, since additional children are seen as additional hands for helping in farming operations.

With some encouragement by the government and with availability of credit, a large number of labor-intensive operations in the long run can be replaced by mechanization and energy-intensive production methods, as is happening in some states of India. Such changes will naturally accelerate the demand for energy in rural occupations. The adoption of new techniques, however, requires integrated development on a large front. For instance, mechanization itself would have inadequate returns unless other inputs were also increased concurrently. Irrigation and high-yielding varieties of seeds and fertilizers are inputs that must be used in combination with mechanized form of agriculture in order to bring about large enough increases in production for an adequate return on private investment. Makhijani has estimated that it requires anywhere from U.S. $100 to U.S. $1,000 per hectare to provide irrigation facilities of different types.[2] By the time required investments in power-generating potential were taken care of, for instance, in tube well irrigation, development costs would total approximately U.S. $320 per hectare. If such a scheme of integrated development was undertaken, even 75 percent of India's cultivated land would require an investment of U.S. $32 billion, a large fraction of the entire GNP of the country.

Makhijani then goes on to mention some examples of how the overall capital costs of development in rural India could be minimized, most of which unfortunately are more in the realm of wishful thinking than practical possibilities.[3] For instance, he suggests that electric motors driving irrigation pumps when not used for irrigation could be applied to grain-milling machines, sugarcane crushers, oil pressers, small-scale industries, and so forth. All these would require managerial and technical skills, apart from various forms of warehousing facilities and so forth, which are currently well beyond the capability of most villages. He also voices the familiar advice of cutting back on automobile (and motorcycle) production to turn out more tractors and farm implements. These are solutions that, in any case, will have only a marginal impact on the overall size of the problem and can only be implemented with a complete reversal of the political system that has evolved in India since independence. Makhijani further mentions that the 6 million automobiles that are junked each year in the United States and sold for scrap at $30 a metric ton could perhaps be used in developing countries for small power plants and generators in villages. It would be naive to imagine that the necessary maintenance support, technical skills, and distribution systems could be established to make such a proposition feasible. In any case, the 6 million junked automobiles in the United States consist of a large percentage of gas guzzlers, which would only aggravate the energy problems of India. In predicting and forecasting developments in the future, it is necessary that one keep in mind the basic skills, expertise, and infrastructure that exist in rural areas to visualize some of the possible changes in activities and production technologies that are likely to take place.

Increases in energy demand are likely to take place not only on account of increased mechanization, which may lead to energy consumption of the order of 1 million to 3 million Btu per hectare per crop, but also because of energy required for irrigation and drainage. Even if the potential of natural resources is utilized efficiently and to the fullest extent, fossil fuel-based energy would be required to a much greater degree than at present. Caution must, however, be exercised in making predictions based on recent developments. For instance, the large-scale exploitation of groundwater reserves in some parts of the country could very well slow down. In fact, in the state of Tamil Nadu, where large concessions, special power tariffs, and recent droughts had led to an almost 500 percent increase in electric pumps installed for irrigation in the last three to four years, the state government is now discouraging further expansion, since the water table has been found to have gone down as a consequence of such large-

ENERGY DEMAND 57

scale exploitation of groundwater resources. The social costs in the form of lower water tables are now being taken into account in evolving integrated water resources management programs. Some of the northern states in the Gangetic Plain have also observed reduction in the availability of groundwater reserves in recent years. Alternatives in the form of major irrigation projects and improved utilization of existing water resources are likely to emerge in the next few years. These will not take place without their attendant effects on demand for energy.

In the above pages, we have dwelt on the developments in the rural sector of the Indian economy, primarily because these are vital not only to the whole phenomenon of economic development of the nation but also to the assessment of the overall demand for energy in the future. Rural activities and consumption patterns will also have a significant effect on movement of demand from noncommercial to commercial sources of energy. In arriving at forecasts for the future, therefore, it is difficult to produce perfectly certain values, and, at best, one can only point to certain tentative directions based on the likely path along which rural society in India as a whole will move forward.

FORECASTS OF DEMAND FOR DIFFERENT ENERGY SOURCES

The reserves of coal available in India will be discussed in the next chapter. At this stage, it is adequate to know that coal is by far one of the most abundant sources of energy available in India. Official policy, therefore, has moved in the direction of trying to promote a greater consumption of coal by substitution—both actual and potential—of other sources of energy. The program for greater dependence on coal has, however, not made great headway and is likely to show results only by the 1980s. There has, nevertheless, been a significant increase both in the production and consumption of coal in the country over the last three years, so that production of coal has risen from 76.8 million metric tons in 1972 to around 100 million metric tons in 1976. The FPC has arrived at forecasts of demand for coal using the end-use method for prediction, and these are shown in Table 3.2.

These figures appear to be highly optimistic. First, it appears the FPC estimated a much larger extent of substitution from other energy sources to coal than was actually realized. Second, on account of a recession in industry during the period 1972-75, the increase in demand based on end use by various industries did not materialize. As a result, the Planning Commission itself has downgraded the target for production of coal from 135 million metric tons in 1978/79 to 124 million metric tons, including a provision of 2.5 million metric tons for export. It appears that even this target may be somewhat unrealistic. The inventory holdings of coal at various stages of the production-distribution cycle are higher than ever before, and availability of wagons for coal handling and movement, normally inadequate in the past, has been highly satisfactory during the last year, thereby indicating that inventories will not be depleted due to shortages in supply.

The Indian Railways, which still uses steam as its major form of traction, and the power generation industry consume 60 percent of the coal produced. The forecasts of demand, therefore, are very sensitive not only to the level of activity in these two sectors but also to the technology employed in the case of the railways, where the mix of technology used is an important determinant. For instance, an accelerated electrification program could reduce the demand for coal considerably, since the average steam locomotive has an overall thermal efficiency of only 4 to 5 percent, as compared to an electric locomotive of about 20 percent (using power generated by a coal-based thermal station). The FPC took into account the increases in fuel efficiency that would be brought about as a result of larger unit sizes and superior efficiency in generating plants of the future (as well as by a reduction in losses in the transmission and distribution system). The demand for coal is also highly sensitive to the growth of the steel industry. India still has a steel-making capacity of well below 10 million metric tons, and in spite of some recent thinking at policy-making levels of achieving a capacity of 75 million metric tons by the turn of the century, the progress made thus far would belie any hopes of this target being achieved.

In our opinion, therefore, the forecasts of coal arrived at by the FPC appear quite unrealistic, and unless strong efforts by government and industry produce spectacular results all over the country, it will be unlikely that the demand for coal in 1990/91 will exceed 250 million metric tons per annum. This would give us an annual rate of increase in production of 6.3 percent. The study by Kirit Parikh has extended forecasts to the period 2000/2001.[4] Parikh has worked out forecasts for demand of 650 million metric tons in the year 2001. His forecasts for 1990/91 are 367 million metric tons; for 1983/84, 218 million metric tons; and for 1978/79, 145.8 million metric tons. These figures, based on a scenario of low population, high growth, and contained urbanization, appear abnormally high. For instance, the assumed steel production is 14.3 million metric tons in 1978/79, 22.3 million metric tons in 1983/84, 36 million metric tons in 1990/91, and 72 million metric tons in 2000/2001. Realizing the long lead time in establishing steel-making capacity, it is already clear that these assumptions starting from 1978/79 itself are unrealistic.

TABLE 3.2

First Estimates of Coal Requirements, End-Use Method,
1978/79, 1983/84, and 1990/91
(millions of metric tons)

Consuming Sector	Expected Requirements		
	1978/79	1983/84	1990/91
Energy use			
Steel plants and coke ovens[a]	32	53	90
Thermal power generation[b]	45	64	118
Transport (railways)	13	11	10
Industries	20	27	50
Brick burning	8	11	20
Domestic soft coke[a]	9	21	30
Export	1	2	3
Collieries own consumption	4	6	9
Total (for energy use)	135	195	330
Nonenergy use			
Fertilizer feedstock	3	6	9
Total coal	135	201	339

[a] In terms of raw coal.
[b] Excluding middlings (5 million metric tons in 1978/79, 12 million metric tons in 1983/84, and 21 million metric tons in 1990/91), which are included under the demand for steel plants and coke ovens.

Source: Report of the Fuel Policy Committee (New Delhi: Controller of Publications, 1974), Table 3.2, p. 14.

The demand for petroleum is another subject on which considerable uncertainty exists. It is likely that official policy in this area, including prices of various petroleum products, will have a significant impact on demand in the future. Again, the FPC's official estimates indicate a total demand, as shown in Table 3.3. The Fifth Five-Year Plan target, which is reflected in the refining capacity that is expected to be set up by 1978/79, is 31.5 million metric tons. The annual consumption of petroleum and petroleum products in 1974/75 was 24.417 million metric tons. On current indication, therefore, it appears that even the anticipated total refinery capacity of 31.5 million metric tons is not likely to be utilized in 1978/79. One of the shortcomings in demand forecasting in this area lies in the fact that the effect of

TABLE 3.3

First Estimates of Demand for Oil Products, End-Use Method,
1978/79, 1983/84, and 1990/91
(millions of metric tons)

Name of Product	1978/79	1983/84	1990/91
Energy sector			
Liquid petroleum gas	0.730	1.200	1.980
Motor gas	2.100	2.550	3.360
Kerosene	3.400	4.400	6.090
Aviation turbine fuel	1.500	2.650	6.120
High speed diesel oil	10.700	15.200	27.910
Light diesel oil	2.050	2.600	3.700
Furnace oil			
Used for power generation and industries	5.500	7.200	10.150
Coastal bunkers	0.230	0.400	0.600
International bunkers	0.175	0.350	0.400
Other	0.330	0.450	0.700
Total	26.715	37.000	61.010
Other than energy sector			
Naphtha	3.470	4.270	5.500
MTO	0.120	0.200	0.300
JBO	0.090	0.100	0.120
Fuel oil for feedstock for fertilizer	1.250	2.000	3.000
Lubes	0.790	1.000	1.300
Bitumen	1.580	2.550	5.000
Petroleum coke	0.312	0.500	1.000
Wax	0.078	0.130	0.250
Total	7.690	10.750	16.470
Grand total	34.405	47.750	77.480

Source: Report of the Fuel Policy Committee (New Delhi: Controller of Publications, 1974), Table 3.3, p. 15.

prices on demand in the long run has not been quantified for petroleum as well as for other sources of energy. In a study carried out by this author, a simulation model was used for arriving at forecasts of demand for the future extending up to the year 2000. Estimates of elasticities of demand were used in this simulation exercise based on the work of Michael Kennedy, who had used data from a large number of countries to estimate this measure.[5] The forecasts arrived at by the author when compared with those of the FPC are as follows:

Forecast of Total Demand for Petroleum (millions of metric tons)	1978/79	1980	1983/84	1985	1990	1990/91
Author's	—	27.20	—	34.71	44.30	—
FPC's	32.20	—	42.58	—	—	66.74

In our view, with a reasonable increase in prices in the future, the demand for petroleum in the country will be much lower than the official estimates of the FPC or the Planning Commission. It must be emphasized again that government policy would have a great impact on regulating demand for these vital commodities. Based on some of the decisions taken in the recent past and the serious impact that continued imports of petroleum would have on the Indian economy, there is every reason to believe that, officially, efforts will be made to curb growth of demand for petroleum.

In arriving at forecasts for demand for electricity, the efforts of the FPC are supplemented annually by the forecasts published by the Central Electricity Authority (CEA). Comment must be made on the methodologies followed by both these agencies in arriving at official forecasts. Whereas the FPC in its report used the end-use method for arriving at electricity demand forecasts, the CEA used three methods and reconciled results obtained by each to arrive at a common forecast. (The three methods used are described in Chapter 6 of this book.) Although these methodologies have served planners and policy makers reasonably well in the past, it is our view that with the changes taking place in relative prices of energy and patterns of consumption, models that can be expanded into greater detail must be employed to arrive at forecasts for the future. These would enable policy makers to test the validity of forecasts produced in response to variations in some of the underlying assumptions. Even though the end-use method used by the CEA in a sense does relate the demand for energy to specific applications, the use of rigid coefficients does not promote the inclusion of changes brought about by relative price changes and adoption of new technologies. Models specifically developing these two concepts must be evolved to suit the needs of the future. This author, in a study carried out for one of the major states in India (Andhra Pradesh), developed a model of this type, which yielded interesting results, particularly in the form of a large number of econometric equations linking demand in various sectors with a large range of economic-demo-

TABLE 3.4

Revised Estimates of Demand for Electricity, End-Use Method,
1978/79, 1983/84, and 1990/91
(billions of kilowatt hours)

Consumption	1978/79	1983/84	1990/91
Major industrial consumption	47.89	80.2	145.0
Other industrial consumption	20.59	25.5	65.0
Total industrial consumption	68.48	115.7	210.0
Domestic consumption	8.68	15.6	25.9
Commercial consumption	5.58	9.8	21.6
Public lighting	1.10	2.0	4.4
Traction (railways)	3.25	5.7	12.6
Irrigation and dewatering	10.00	14.2	40.5
Total consumption	97.09	163.0	315.0
Losses and auxiliary consumption	22.77	35.8	70.0
Total generation required	119.86	198.8	385.0

Source: Report of the Fuel Policy Committee (New Delhi: Controller of Publications, 1974), Table 3.4, p. 15.

graphic variables. The forecasts of the FPC for demand for electricity are given in Table 3.4. As against this, the forecasts of the CEA are given in Table 3.5. In the study carried out by this author, it was found that the estimates of the CEA were higher than those obtained by us. If our results for the state of Andhra Pradesh can be taken as representative of the whole country, the forecasts shown in Table 3.5 above would turn out to be much higher than reasonable. The model developed by us and the results obtained therefrom are explained in much greater detail in Chapter 6, which deals with the electrical energy sector in India.

In the FPC report, a number of pages are devoted to forecasts of demand for commercial sources of energy, but less than a page is provided for estimates of demand for noncommercial fuels. This is perhaps as much a reflection on the prevailing neglect of noncommercial energy studies as of the tremendous difficulties and paucity in availability of organized data related to this form of energy. The biggest difficulty in arriving at any quantitative conclusions on noncommercial energy demand lies in the fact that there is no possibility of

TABLE 3.5

Total Electrical Energy Requirements, 1978/79–1983/84 and 1990/91
(millions of kilowatt hours)

Region	1978/79	1979/80	1980/81	1981/82	1982/83	1983/84	1990/91
Northern	34,078.0	37,765.0	41,852.0	46,402.0	51,413.0	57,000.0	110,910.0
Western	37,223.0	40,732.0	44,461.0	48,378.0	52,698.0	57,350.0	102,480.0
Southern	31,188.0	34,204.0	37,605.0	41,398.0	45,573.0	50,150.0	94,900.0
Eastern	23,051.0	25,350.0	27,880.0	30,650.0	33,800.0	37,210.0	70,400.0
Northeastern	1,457.0	1,783.0	2,152.0	2,495.0	2,896.0	3,350.0	9,307.0
Andaman and Nicobar Islands	10.0	11.5	13.2	15.2	17.5	20.0	50.0
Lakshadweep Islands	1.0	1.2	1.3	1.5	1.8	2.0	5.0
Total	127,008.0	139,847.7	153,965.5	169,340.7	186,399.3	205,082.0	388,052.0
Nonutilities	6,500.0	7,100.0	7,750.0	8,450.0	9,200.0	10,000.0	19,000.0
Grand total	133,508.0	146,947.7	161,715.5	177,790.7	195,599.3	215,082.0	407,052.0

Note: Energy requirements indicated above represent the net generation requirement and do not include the consumption in power station auxiliaries.

Source: Government of India, Central Electricity Authority, Ninth Annual Power Survey (New Delhi, 1975).

direct measurement of noncommercial usage, since these energy sources do not move through established distribution channels and markets. The methodology used for arriving at future estimates relies on using the value of average per capita consumption in the domestic sector in both urban and rural households and then working out the total requirement of energy based on the distribution of urban and rural populations of the country. This is arrived at by first assessing the availability of commercial fuels and, then, treating the balance as the amount that will have to be made good by noncommercial sources. This methodology is hardly rigorous and completely ignores the effect of prices on domestic consumption. Commercial sources of fuel, even though available, may well be beyond the reach of the poor, who find it cheaper in terms of opportunity costs to collect various forms of noncommercial fuels. As it is, of course, noncommercial fuels are hardly used for any other type of consumption, except domestic. It is not unlikely, nevertheless, that in the future, waste recycling plants, biogas plants, and so forth could produce energy on a relatively large scale, which could be utilized for such activities as irrigation, heating for industrial applications, and so forth. Table 3.6 provides estimates of demand for noncommercial fuels for three selected areas in the future. These figures reflect an aggregate decrease in demand for noncommercial fuels in the future. This is predicated on an increase in the share of commercial energy as a percentage of the total. The specific figures mentioned in the table are based on a consumption of 20.1 percent of commercial energy in 1978/79 and 39.9 percent in 1990/91 in the domestic sector. These figures must also be regarded with some skepticism. For instance, the demand for firewood has been increasing slowly, at the rate of about 2 million to 3 million metric tons a year. There is little reason to believe that unless major socioeconomic changes take place, this rate of increase will slow down (it might actually increase). Decisions on consumption of firewood, as for other noncommercial sources, are based on relative prices of commercial and noncommercial fuels. The FPC assumes demand for firewood to grow up to 1983/84 beyond which it is expected to decline. Little has been said to justify this assumption. The total availability from recorded fellings in the forest under the control of the government is, according to the FPC, likely to go up to 35 million metric tons. The unrecorded removal of wood is likely to remain around the present level of 90 million metric tons. We believe that the figures obtainable through official sources, as used by the FPC, grossly understate the extent to which forests are contributing to the use of firewood.

    The total output of dry cow dung based on the population of livestock in this country is a little less than 200 million metric tons. Almost one-third of this is consumed as fuel in the country. Cow dung

TABLE 3.6

Estimates of Demand for Noncommercial Fuels,
1978/79, 1983/84, and 1990/91

Fuels	1978/79		1983/84		1990/91	
	mmt	mmtce	mmt	mmtce	mmt	mmtce
Firewood and charcoal	132	125	131	124	122	116
Dung cake (dry)	65	26	65	26	53	21
Vegetable waste	46	44	46	44	46	44
Total	243	195	242	194	221	181

Note: mmt = million metric tons; mmtce = million metric tons coal equivalent.

Source: Report of the Fuel Policy Committee (New Delhi: Controller of Publications, 1974), Table 3.6, p. 20.

has a competing use as fertilizer, and most estimates indicate that the value of marginal product from cow dung as fertilizer is much higher than that of cow dung as fuel. A new method that is gradually finding acceptance in the country is the use of gobar gas (biogas) plants, with the use of cow dung as feedstock. These plants not only provide cow dung gas for use as fuel but also produce large quantities of fertilizers through a relatively efficient process. Undoubtedly, installation of gobar gas plants on a large scale will result in a reduction in demand for gobar, if the use of its products is confined to present uses only. However, it is entirely possible that when cow dung gas and good-quality manure is produced in this manner, a large number of other forms of consumption will be conceived of that will result in an increased demand for cow dung. To assume that this increased demand may result in a corresponding decrease in other forms of commercial energy may also not be realistic, since the major uses of gobar gas are likely to be in the rural areas and since these are not currently consumers of commercial forms of energy.

Vegetable waste is also used on a fairly large scale in some parts of the country. Bagasse is one known form of vegetable waste used for energy production. But it also has applications as an input for the manufacture of paper, and this particular application has been pushing up its price in the market. With the opportunity cost of using the bagasse as fuel going up, it is likely that its consumption as a fuel will go down, but this would be attendant on an increase in demand for industrial applications. For this reason, the FPC has assumed the level of usage of vegetable waste will remain at the same level in 1990/91 as it is expected to be in 1983/84.

Different forecasts for energy demand have been produced in a number of studies recently. The study by Parikh extended the work of the FPC and produced some additional refinements.[6] It attempted to arrive at forecasts for the year 2000/2001. In doing this, the basic forecasts of the FPC up to 1990/91 were retained, and assumptions on economic growth, urbanization, and population growth were made for the period 1990/91–2000/2001.

As an illustration, in Table 3.7, we have reproduced estimated requirements of fuels up to the year 2000/2001, as arrived at by Parikh on the assumptions of low population, high growth, and contained urbanization. These forecasts, particularly with respect to commercial fuels, appear to be unreasonably high and, correspondingly, the estimated requirements for noncommercial fuels unreasonably low. Parikh has not quite explained how the massive substitution in favor of commercial fuels in rural areas will be brought about. Within the commercial fuels themselves, some figures have been provided for interfuel substitution in favor of coal and electricity from consumption of oil products. As mentioned by him, the potential for substitution of oil products is greatest in the transport sector. Rail transport is relatively less intensive in fuel consumption, and it is possible that a fairly substantial amount of interfuel substitution can be brought about by making rail transport more attractive for freight traffic, as against road transport. The basis for arriving at requirements of noncommercial fuels is the FPC's assumption of 0.40 and 0.38 metric tons of coal replacement per capita per year in urban and rural areas, respectively.

In making these estimates, the effect of income elasticity on future demand has not been taken into account. In a study carried out by the National Council for Applied Economic Research, it was found that the income elasticity of domestic energy consumption lies probably between 0.18 and 0.20.[7] These estimates of elasticity are based on very inadequate data and can at best be taken as a crude indicator of the income effect on demand for noncommercial fuels. Nevertheless, with projected increase in rural income in the future, the figure of per capita consumption assumed in the study could very well prove to be low. Another factor that has not been accounted for in these forecasts is the effect of relative prices on demand for noncommercial fuels. There is already serious thinking on increasing prices of electricity for rural applications, such as tube well irrigation and so forth. Coupled with this is the present policy to price kerosene at levels almost on a par with diesel oil in order to restrict the possibility of adulteration of the latter. With possible increases in the price of

TABLE 3.7

Estimated Requirements of Fuels, 1978/79, 1983/84, 1990/91, and 2000/2001, and Actual Consumption, 1970/71
(in original units)

Years	Coal[a] (10⁶ metric tons)	Oil[a] (10⁶ metric tons)	Electricity (10⁹ kilo-watt hours)	Firewood (10⁶ metric tons)	Animal Dung (10⁶ metric tons)	Agricultural Wastes (10⁶ metric tons)
1970/71	66	18	56	123	67	38
Case a[b]						
1978/79	135	34	120	132	65	46
1983/84	201	48	200	131	65	46
1990/91	339	77	385	122	53	46
2000/2001	600	145	670	89	40	46
Case b[c]						
1978/79	146	30	128	132	65	46
1983/84	218	39	211	131	65	46
1990/91	365	57	398	122	53	46
2000/2001	650	97	700	89	40	46

[a]Including coal and oil used in generating electricity and nonenergy sector.
[b]Crude oil price U.S. $5 per barrel.
[c]Crude oil price U.S. $10 to U.S. $12 per barrel.

Source: Kirit Parikh, Second India Studies: Energy (Delhi: Macmillan Company of India, 1976), Table 4.2, p. 50.

kerosene, there could be a substitution effect in favor of noncommercial fuels. Therefore, the result would be that per capita consumption of noncommercial fuels in the rural sector would actually go up in the years to come. Parikh's estimate can, therefore, not be taken as very reliable, unless some of these income and price effects are studied in depth.

Energy modeling and forecasting in most economies is a very complex and difficult subject. The problem is certainly complicated in developing societies, where estimates, particularly with respect to noncommercial sources of fuel, are very difficult to come by. The result is that studies in the past that have attempted to arrive at forecasts of demand have really arrived at forecasts of requirements based on rigid coefficients and constant prices and incomes. The economic realities of energy demand, however, are quite to the contrary. Particularly with the changes taking place in relative prices of commercial fuels and a likelihood of greater incomes in the rural sector, forecasts will have to be produced by more rigorous efforts. Scholars of energy policy will, therefore, need to focus on more reliable methods of forecasting before any set of policy alternatives can be fully appreciated and evaluated.

NOTES

1. Arjun Makhijani, Energy and Agriculture in the Third World (Cambridge, Mass.: Ballinger, 1975), pp. 72-73.
2. Ibid., pp. 76-77.
3. Ibid., pp. 78-79.
4. Kirit Parikh, Second India Studies: Energy (Delhi: Macmillan Company of India, 1976), Table 4.3, p. 52.
5. Michael Kennedy, "An Economic Model of the World Oil Market," The Bell Journal of Economics and Management Science 5, no. 2 (Autumn 1974): 540-77.
6. Parikh, Energy, pp. 88-114.
7. National Council of Applied Economic Research, Domestic Fuels in Rural India (New Delhi: National Council of Applied Economic Research, 1965), p. 45.

CHAPTER

4

ENERGY RESOURCES

Various studies in the past have made an assessment of energy resources in India. Foremost among these are the report of the Energy Survey of India Committee,[1] and various studies carried out on a regional basis by such agencies as the National Council for Applied Economic Research (NCAER).[2] A detailed and integrated view of the entire scene becomes necessary, mainly as a result of the energy crisis of 1973-74. India, in effect, had no energy policy prior to the events of 1973-74; a systematic assessment of energy resources was carried out first by the FPC.[3]

India is endowed with fairly good resources for the production of energy. These include vast reserves of coal, promising structures for oil and natural gas, extensive hydroelectric potential, and nuclear fissile material. As mentioned in the previous chapter, the availability of a vast amount of noncommercial fuels, such as firewood, cow dung, and vegetable waste, are important in assessing the overall resources picture. The geographic situation of India makes it a very favorable country for the exploitation of solar energy on a large scale. This, however, cannot be considered as a tangible resource until such time as technical developments permit commercial utilization of solar energy.

PETROLEUM

India's place (as one of the countries affected adversely by the oil price hike of 1973-74) merits a detailed examination being made of its oil and natural gas resources. The efforts of the Oil and Natural Gas Commission (ONGC) in recent years has raised hopes about the existence of fairly substantial deposits of oil and natural gas in the country, both onshore and offshore. There is no doubt that India's efforts to become self-sufficient were not as effective as they might have been, both before and after independence. It would be useful to explore the history of the oil industry in India to arrive at reasons for this lag and to see how indigenous development has now taken the country closer to self-reliance in this vital field.

The entire oil sector in the country prior to independence was dominated by foreign-owned corporations. These included Burmah-Shell, Standard Vacuum Oil Company, and Caltex. It is wrong to assume that the importance of oil imports to India has emerged only recently. As far back as 1954, when India's consumption was approximately 31 million barrels, the cost of imported petroleum and petroleum products amounted to approximately U.S. $200 million, or 15 percent of the country's total import bill. Interestingly, petroleum imports were almost three-fourths as large as food imports. This import bill was reduced by action on the part of the Indian government, which led to the three oil giants that were operating in the country setting up three refineries in the 1950s. This did not, however, lead to any change in product prices in India, since the multinationals set prices equal to product prices in the Middle East (plus freight and handling charges and so forth).

The import of crude oil and the prices paid by the multinationals remained in their own hands. By 1960, however, the Indian government succeeded to a large extent in getting the three major companies to import crude oil rather than refinery products, thus resulting in foreign exchange savings. With increased industrialization and all-around growth, the total demand for oil continued to grow rapidly.

The government was, therefore, subjected to a lot of pressure due to increasing foreign exchange requirements for oil imports. In response to this challenge, attempts were made to step up crude oil exploration within India and, at the same time, to try to reduce the import prices for crude oil by diversifying the sources of supply and putting pressure on the major oil companies to pay lower prices to their affiliates overseas. This led to a barter agreement being arrived at between the Soviet Union and India (India would import oil and would export certain commodities to the Soviet Union). Oil companies were, therefore, asked to import large quantities of oil from the Soviet Union at a considerably lower price than existing prices. Specifically, the Soviet Union offered to supply some 18 million barrels of crude oil per year at a price of approximately U.S. $1.81 per barrel, which was almost 15 percent cheaper than the oil imported by the multinationals. This deal, as offered by the Soviet Union, led to a great deal of political argument and debate. Even though it placed the multinationals in an awkward position, they resisted the import of Soviet oil, possibly under the influence of their own governments.

Since they controlled the refining capacity available in India at that stage, they came into disagreement with the Indian government. It was obvious that the Soviet deal would not only be politically embarrassing but would also cut into the large profits that the oil companies were making by importing from their own affiliates. They were also worried that in those countries where similar organizations existed, the Soviet deal with India could lead to similar actions.

With the backing of the U.S. and British governments, and due to some division within the Indian government itself, the multinationals were able to force a rejection of the Soviet offer. However, this led to considerable friction between the Indian government and the multinational oil companies. This controversy led to a demand for nationalization of the oil industry, voiced by the left wing of the Indian government, and a serious effort was made to develop an integrated oil policy for India. A committee was set up in the early 1960s, known as the Damle Committee, to cover three aspects of oil policy, namely, crude oil supplies for Indian refineries, price of refined products, and allowable expenses for marketing and distributing products in India. This committee found that crude oil prices were being sharply discounted by many private companies in other parts of the world. It, therefore, recommended that the companies operating in India buy crude oil from lowest-price suppliers or ask their own suppliers to cut prices to reach the lowest existing levels, as well as eliminate intermediate purchasing companies that acted as middlemen between suppliers in other countries and the buyers in India. The oil companies did not find it in their interest to cooperate with the government by accepting these recommendations.

It became apparent to the Indian government, in light of their problems with handling the multinationals, that the stranglehold the multinationals had over the oil sector of the country had to be broken. This was attempted in two stages. The first was to set up refining capacity under national control in addition to the existing capacity controlled by the multinationals. Again, after some debate within the Indian government itself, it was decided to set up refineries in the public sector. The lack of response from Western countries in providing the technical and economic assistance required for this purpose led to the Soviet Union and Romania entering into the vacuum.

Two refineries were set up in the eastern part of India and one in western India to make use of indigenously produced oil in that region. By 1966, the three government refineries had a total capacity of 51 million barrels per year. This capacity, however, was not utilized fully in the initial years of production, mainly because of various technical snags and difficulties in running them. As such, the gap had to be made good by permitting the various foreign companies to expand their capacities in the existing refineries. Between 1957 and 1961, private sector refinery capacity increased from 4.33 million to 6.05 million tons. The conflict between foreign oil companies and the Indian government was reflected within the Indian Congress Party also. (A large measure of the credit for efforts at self-reliance in the oil sector go to K. D. Malaviya, who at different stages was responsible for the Petroleum Ministry in the Indian government. He was, however, identified with the leftist section of the Congress Party, and his actions were often opposed and questioned by the more moderate and right-wing sections within the party.)

Part of the reason why strong action was not taken against the multinationals by the Indian government lies in the fact that the Russians were not willing to sell large quantities of refined products to India without some expenditure in foreign exchange. They were also not willing to provide technical and economic assistance for large-scale expansion of refinery capacity during the 1960s. As a result, therefore, the course adopted by the Indian government was a form of compromise, where it had to live with the reality of the foreign companies importing crude oil from their own sources. The Indian government could at best counter their power by developing its own capacity for crude oil production and refining.

Up to the early 1950s, the Burmah-Shell group was the only major company engaged in oil exploration in India. Some sporadic efforts were made by other groups to locate reserves of oil or natural gas, but these did not yield significant results. Standard Vacuum, for instance, entered into a partnership agreement with the Indian government and drilled ten test wells but found no oil or gas; the entire project was closed down in 1960, after considerable expenditure.

In 1956, an Indian delegation went abroad to various countries to study the oil industry in other parts of the world and to determine how and what assistance could be obtained in India's own efforts. The countries visited were the Soviet Union, Romania, Great Britain, the United States, and some parts of Latin America. Consequent to this tour, Malaviya, who headed the delegation, spearheaded the establishment of ONGC in 1956. Even though technical help for the earlier activities of ONGC came from Western countries, as well as the Soviet Union, it was the latter that provided the bulk of funds for importing drilling equipment and for training facilities. As a result, therefore, it took a dominant position in assisting the activities of ONGC. In 1958, the first major event for national control for oil exploration in India occurred with the discovery of the Cambay gas field. Other discoveries followed in the next three or four years. During the first ten years of the existence of ONGC, the Indian government spent close to U.S. $4 million on exploration and production. Michael Tanzer estimated a total finding cost of about U.S. $0.35 per barrel during that period.[4] Such estimates, however, are not reliable, since

ENERGY RESOURCES

most data pertaining to oil and natural gas exploration in India are classified.

The foregoing narration has been presented to emphasize the disastrous effects the role of multinational corporations has had on India's own efforts to be self-reliant in the petroleum sector. This is not so much a sad commentary on the behavior of multinationals as on the objectives on which their operations were organized. Acknowledging the fact that most of the multinational oil companies were already in charge of low-cost crude oil, which they could produce in large quantities in the Middle East, there would naturally be little incentive for them to use or produce oil from any other source. That the oil available indigenously in India would have been beneficial for Indian consumption in economic terms is borne out by the fact that the price of oil supplied by ONGC to Indian refineries has consistently been lower than the price of imported oil obtained through the major foreign companies. This wide disparity has become wider still after the increase of international oil prices in the wake of the 1973-74 upheaval.

Apart from ONGC, two other companies are engaged in oil exploration and production (on a limited scale). These are Oil India (which until recently was under foreign control) and Assam Oil Company. The major producer of oil indigenously has, however, been ONGC, and the expansion that has taken place in its operations is fairly impressive, considering the fact that it has a history that extends back only 20 years, as compared to the multinationals, who have been in the business for at least four times this period.

The cumulative production of crude oil until the end of fiscal year 1974/75 was 35.78 million metric tons (ONGC's production by years is shown in Table 4.1). With the success met by its offshore ventures, ONGC is currently planning to step up its level of production to about 20 million metric tons by the year 1980/81. Of this, about 10 million metric tons is expected to come from the recently found Bombay High oil structure.

The increase in oil prices that is taking place internationally has provided a fillip to ONGC's efforts in recent years. Recognizing the importance of exploration activities, the government of India has allowed for an expenditure of 1,056 crores of rupees, as compared to the 420 crores of rupees provided in the draft Fifth Five-Year Plan. With this expenditure, it is expected that the 1978/79 production target for indigenous crude oil of 14.18 million metric tons will be met. At the same time, refining capacity is also being stepped up and, by the end of the fifth plan, that is, 1978/79, the total refining capacity in the country is expected to reach 31.5 million metric tons per year. Even though in terms of international comparisons, India's efforts in self-sufficiency do not compare with production levels obtaining in other

TABLE 4.1

Oil and Natural Gas Commission's Output of Crude Oil,
1967/68–1974/75
(millions of metric tons)

Year	Crude Oil Production
1967/68	2.8
1968/69	3.09
1969/70	3.64
1970/71	3.61
1971/72	4.01
1972/73	4.11
1973/74	4.03
1974/75	4.52

Source: Oil and Natural Gas Commission, Annual Report 1974-75 (Dehra Dun, 1975).

countries, it is certainly creditable that within a short period of 20 years, much has been achieved, particularly in a region where oil does not exist in abundance. The FPC estimated that up to 1971, only 4 percent of the likely oil-bearing regions of the country had been explored. A massive program for exploration is, therefore, likely to lead to substantial benefits, given a certain amount of perseverance and financial backing by the government.

Immediate gains from exploration efforts are likely to come from offshore areas, particularly the Bombay High region, which is located 110 to 180 kilometers northwest of Bombay in the Arabian Sea. Other structures in that region have also proved promising. What is heartening about the Bombay High discovery is the fact that oil has been found for the first time in India in the limestone layers of the structure under the seabed. Hitherto, oil had been found in India only in the onshore sandstone areas of Assam and Gujarat. Limestone deposits are known to be richer producers of oil, as found, for instance, in the Middle East. Also, offshore wells are generally more productive. Whereas the average production of India's onshore wells is 180 barrels a day per well, the daily flow rate of offshore Bombay High wells was 2,000 to 3,000 barrels during production tests without acidization. With acidization and other recovery techniques, production levels are likely to be increased considerably. The current production of crude oil in India is at a level of over 8 million metric tons. Of this, 3 million metric tons are produced by Oil India, and the Assam Oil Company produces approximately 70,000 metric tons. The balance comes from the oil fields controlled by ONGC.

India has approximately 390,000 square kilometers of continental shelf of sedimentary basins. The history of offshore exploration in India started in 1963, when a survey of the Gulf of Cambay area was carried out. This led to the discovery of the Aliabet-West and Aliabet-East and Tapti structures. The vessel originally chartered by ONGC was found unsuitable for exploration in deeper regions. Hence, a Soviet seismic survey ship was obtained on contract basis for carrying out surveys from 1964 to 1966. This ship surveyed the Gulf of Kutch, sone parts of the Gulf of Cambay, and the Arabian Sea. It also mapped out areas along the coast of Kerala, the Gulf of Mannar, the Palk Straits, and the Coromandel Coast, as well as some parts of the Bay of Bengal. The survey gave indications of promising structures for the southern part of the Gulf of Cambay and the Bombay High region, as well as the Bassein, Tarapore, and Diu structures in the Arabian Sea.

Offshore drilling was started by ONGC drilling the first well in the Aliabet Bed West structure in 1970, making use of a fixed platform. In order to give momentum to offshore drilling, consultants were employed from other countries, and an Indo-Japanese collaboration was established after the Indian prime minister's visit to Japan in 1970. India's continental shelf comprises mainly ten basins, with average water depths of 80 kilometers. These basins, with approximate areas, are Bombay High (75,000 square kilometers), Kutch (28,000 square kilometers), Saurashtra (32,000 square kilometers), Kerala-A (41,160 square kilometers), Kerala-B (40,320 square kilometers), Cauvery (26,320 square kilometers), Godavari (32,760 square kilometers), Mahanadi (12,040 square kilometers), Bengal (28,280 square kilometers), and Andaman (35,840 square kilometers).

Bombay High has been exclusively earmarked for oil exploration by ONGC, but the remaining nine basins are likely to be explored with foreign collaboration on a production-sharing basis. This policy, originally designed to hasten the goal of self-reliance, may undergo some revision, because in recent months there has been a realization that foreign collaborators who were offered production-sharing terms have not responded favorably in their efforts.

The potential for possible reserves in offshore areas can be gauged by the estimated potential of the Bombay High structure. According to present estimates, it is difficult to arrive at exact figures, but reserves in the Bombay High region are likely to be well in excess of 1 billion barrels. This may be sufficient to sustain a production rate of 10 million metric tons per annum for almost 15 to 20 years. To achieve a production rate of 10 million metric tons, as mentioned, 500 wells may have to be drilled in the area at a total expenditure of about 500 crores of rupees. Whereas the cost of production from offshore regions is relatively low, transportation costs are generally very much higher than for onshore areas. This, therefore, limits serious efforts at exploration to regions in close proximity to the coastline and to those areas where the sea is relatively calm all year round. Estimates of current reserves of crude and natural gas are not very accurate. The Indian Petroleum and Chemical Statistics, in 1972, gave the crude oil reserves of India as 117.84 million metric tons and of natural gas as 62.48 million cubic meters. With the existing offshore reserves that have been discovered, these figures with respect to oil are likely to be in excess of approximately 250 million metric tons, based on current indications. The ONGC has charted the potential of different parts of the country for oil exploration. These are given in Table 4.2.

Mention must also be made of the foreign operations of ONGC. These have been undertaken through a subsidiary of ONGC, Hydrocarbons India. Present programs of ONGC extend to Iran and Iraq, but there is little hope of finding and sharing oil on an appreciable scale in either country. These foreign operations testify to the technical abilities that ONGC has been able to build up in a short period of time. India, as a whole, is in a unique position with respect to technical manpower: it possesses the third-largest scientific and technical manpower in the world (after the United States and the Soviet Union).

Some of the technical and managerial achievements of ONGC are represented in Figures 4.1 and 4.2. Figure 4.1 indicates the total length of drilling achieved by the company in 1,000 meters. Superimposed on this is also the commercial speed and cycle speed achieved in each year. Figure 4.2 shows the number of wells worked out by work over rigs in operation. The commission also has a very large number of trained technologists and geologists whose know-how and skill is of a high order. Overall, therefore, the potential resources that India possesses in the petroleum field are promising and may lead to self-sufficiency in the long run.

COAL

Coal is by far the largest-known energy resource existing in the country. The resources of coal in India have been assessed by the Geological Survey of India (by regional mapping and various forms of drilling operations). A lot of work was also done by the (now defunct) National Coal Development Corporation and some state governments in the country. Detailed estimates of reserves are given in Table 4.3. The deposits of coal that exist in India are known to be

TABLE 4.2

Sedimentary Areas, by Category

Sedimentary Basins, Including Offshore Parts	Total Sedimentary Area, Including Offshore Parts (thousands of square kilometers)	Remarks
Cambay, large part of Assam-Arakan	300	High prospects—these are areas with thick sections of marine sediments of relatively younger age and are known to contain commercial oil/gas
Small part of Assam-Arakan, Cauvery, West Bengal, Jaisalmer, Tripura, Cachar, and Andaman Nicobar	250	Medium prospects—these are areas with thick sections of marine sediments of relatively younger age, but so far have no known commercial oil/gas fields
Ganga Valley, Punjab, Godavari-Krishna, Damodar Graben, Mahanadi Graben, Narmada, Kutch, Saurashtra, Kerala, Laccadive, and Palar	670	Fair prospects—these are areas with thick sections of sedimentary rocks of older ages, either wholly or mostly nonmarine or have some other unfavorable geological feature
Cuddapah, Kaladgi Bhima, Chatisgarh Bastar, Vindhyan Jodhpur, South Rewa Satpura, Sancher-Barmer	450	Low prospects—these are areas with sedimentary rocks of extremely old age

Source: Oil and Natural Gas Commission, Report of the Fuel Policy Committee (New Delhi: Controller of Publications, 1974), Table 4.3, p. 25.

FIGURE 4.1

Drilling Performance

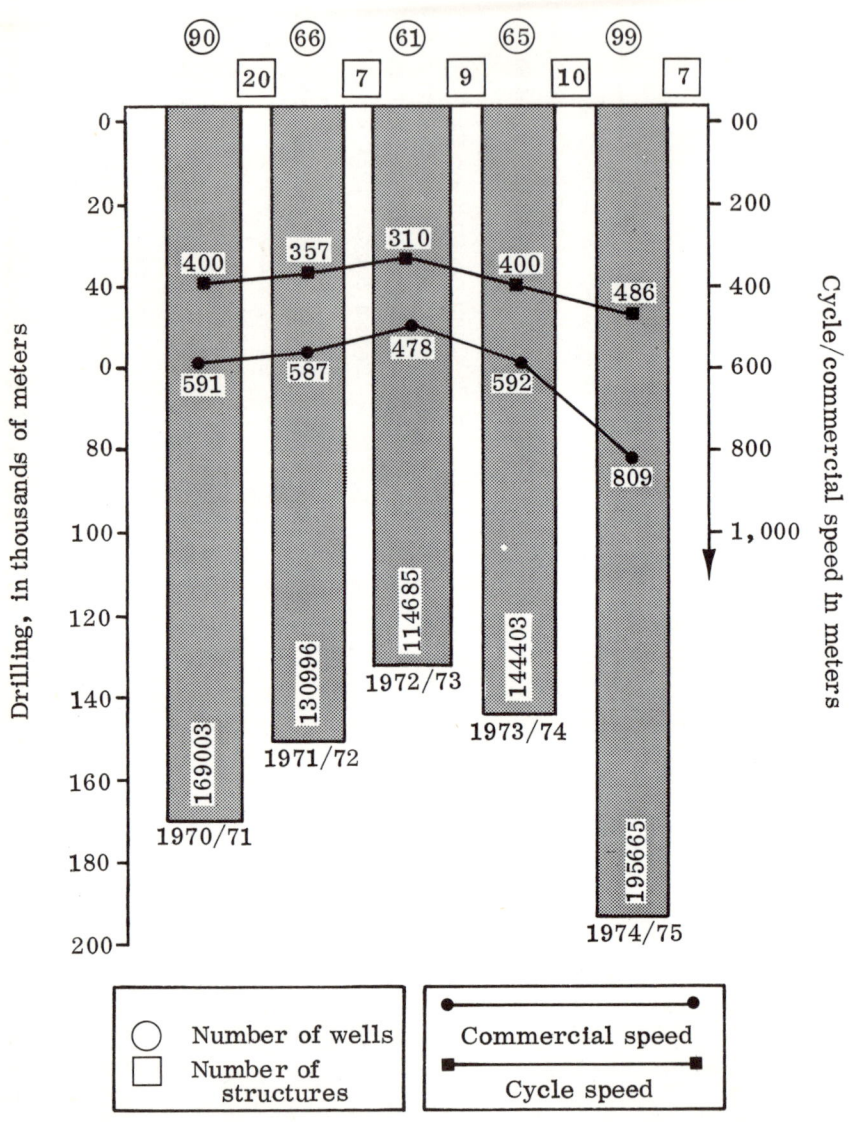

Source: Oil and Natural Gas Commission, _Annual Report 1974-75_ (Dehra Dun, 1975).

FIGURE 4.2

Number of Wells Worked Over by Work Over Rigs

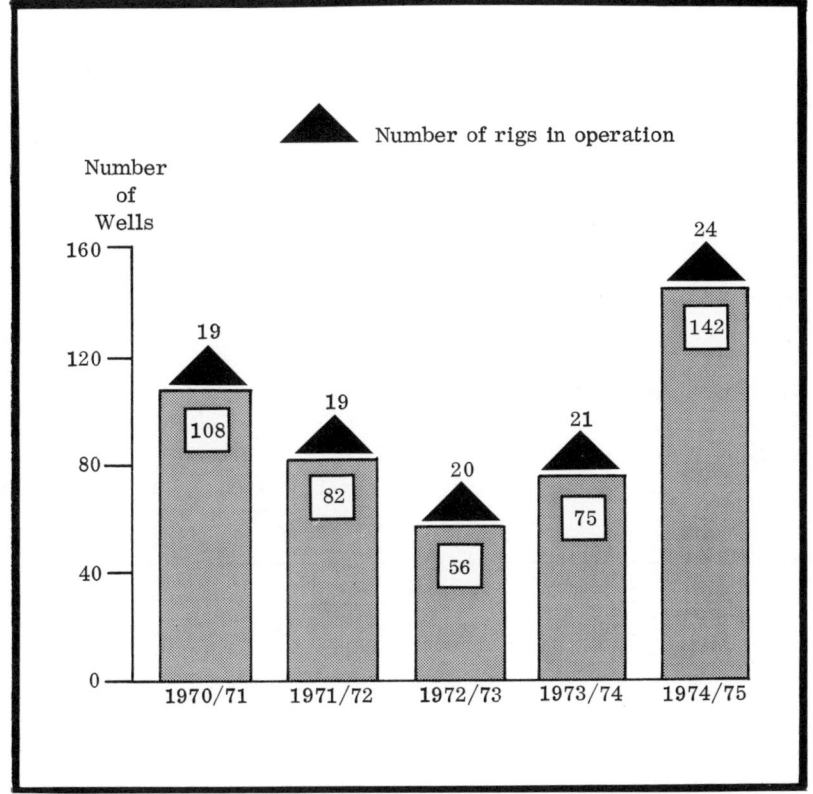

Source: Oil and Natural Gas Commission, Annual Report 1974-75 (Dehra Dun, 1975).

largely of poor quality. Of the total metallurgical coal reserves (approximately 20,000 million metric tons), over 9,000 million metric tons have been proved. Coking coal reserves can be used for the production of metallurgical coal; medium-coking, semicoking, and weakly coking coal must be used in combination with prime-coking coal for preparing metallurgical coke. It, therefore, becomes essential for metallurgical processes to use at least some part of the prime-coking coal available in the country. The resources of prime-coking coal are rather limited, being just over 5,000 million metric tons in the aggregate. The FPC has estimated that along with the estimated availability of medium and blending coal, the steel industry can be

TABLE 4.3

Reserves of Coal
(millions of metric tons)

Type of Coal	Total Gross Reserves	Proved Reserves	Indicated Reserves	Inferred Reserves
Coking coal				
Prime-coking coal	5,650	3,650	1,540	460
Medium-coking coal	9,431	3,850	4,309	1,272
Semi- to weakly coking coal	5,073	1,559	2,600	914
Total-coking coal	20,154	9,059	8,449	2,646
Noncoking coal				
Noncoking coal	59,968	12,306	22,310	26,180
Tertiary coal	288			
Lignite				
Lignite	2,025	1,795	202	28
Grand total	82,975	23,160	30,961	28,854

Source: Government of India, Task Force on Coal and Lignite, *Report of the Fuel Policy Committee* (New Delhi: Controller of Publications, 1974).

supplied with its requirements from existing coal reserves only for about 40 years.[5] Conservation measures are, therefore, of great importance in the steel industry. Another problem with the reserves of coking coal available lies in the fact that a large part of it would have to be beneficiated before it could be used for metallurgical coke. Existing technologies cannot otherwise use the high-ash content variety of coke directly in blast furnaces. There are also possibilities of beneficiating noncoking coal for use in the steel industry.

Reserves of noncoking coal and lignite are relatively abundant. Estimated reserves are in the region of over 60,000 million metric tons. This type of coal is likely to be used in larger quantities with the shift in emphasis to thermal power generation that is taking place in the power sector of the country. Even with pessimistic estimates, it seems that existing reserves of noncoking coal would last close to another 150 years. It is also interesting to observe that most of the coal reserves that are known today are either located within, or are extensions of, the deposit areas that were known about 30 years ago.

ENERGY RESOURCES   81

It is highly likely that other seams exist in the country that may be discovered in the future, thereby increasing the total reserves known to be available.

Another feature of Indian coal is the unusually high-ash content. Of good-quality coals, which are classified as containing not more than 19 percent of ash plus moisture, only 460 million metric tons exist. The regional dispersal of total coal deposits in the country is shown in Table 4.4. The development of the coal industry in the country has been hampered to a large extent by the direct import of technologies from Western countries, which had abundant access to supplies of petroleum and thus did not stimulate demand for coal, which may have led to modernization of the coal industry. This feature is part of a worldwide trend that has been observed, even in such countries as the United Kingdom and West Germany, where the existence of large deposits of coal did not appreciably affect the increasing dependence on oil as a source of energy in industrial and agricultural activities. The result has been that most of the coal mines in the country until recently have suffered from the use of outdated equipment, the existence of hazardous working conditions, and poor financial health. Targets for production set up by the government in accordance with the various five-year plans were revised downward repeatedly. This happened not so much because of the inability of coal mines to supply the quantities targeted but (primarily) because expected demand was not realized. Even in the Fifth Five-Year Plan, the original target for annual coal production was 135 million metric tons by 1978/79. This has now been revised to a figure of 124 million metric tons, including 2.5 million metric tons for export. The slow increase in India's demand for coal, combined with problems in transportation in the past, led to accumulations of large stocks of coal at the pitheads. Low levels of demand also resulted from falling real prices of petroleum in the period 1950/51-1970/71, as well as from the poor progress of planned projects in industry during this period.

A large component of the costs of using coal is related to the high-ash content of Indian coals. These take the form of extensive ash removal facilities in coal-burning plants and the additional unprofitable cost of transporting coal, of which a large percentage exists in the form of ash, which cannot be used to provide energy. As a national policy, it was decided to set up coal washeries to reduce the ash content of Indian coals. Most of these in the past were established by steel plants to reduce the intake of ash in blast furnaces. Only about 4 million metric tons per year of the total capacity established for washeries was intended to be used for noncoking coal, and even this has been poorly utilized. The actual utilization of washeries as a whole is about 60 percent of their rated capacity, and only very recent efforts have improved on this poor performance. By 1975/76,

TABLE 4.4

Regional Availability of Coal Reserves, by Category
(millions of metric tons)

Region	Proved	Indicated	Inferred	Total
Noncoking coal				
Northeastern region	139	291	398	828
Eastern region				
West Bengal	3,071	522	8,848	17,139
Bihar	2,603	8,647	6,612	17,862
Orissa	895	2,411	1,810	5,186
Central region, Madhya Pradesh	4,142	3,864	7,168	15,174
Western region, Maharashtra	478	800	1,344	2,622
Southern region, Andhra Pradesh	978	1,077	—	2,055
Total	12,306	22,310	26,180	60,796
Coking coal				
Eastern region				
West Bengal	967	775	738	2,480
Bihar	7,987	7,470	1,908	17,365
Central region				
Madhya Pradesh/Uttar Pradesh	105	204	—	309
Total	9,059	8,449	2,464	20,154
Lignite				
Western region, Gujarat	78	—	—	78
Southern region, Tamil Nadu	1,717	202	—	1,919
Northern region				
Rajasthan	—	—	20	20
Kashmir	—	—	8	8
Total	1,795	202	28	2,025
Grand total	23,160	30,961	28,854	82,975

Source: Report of the Task Force on Coal and Lignite, April 1972, from Report of the Fuel Policy Committee (New Delhi: Government of India, Controller of Publications, 1974), Table 4.2, p. 24.

ENERGY RESOURCES

the capacity for washing was less than one-third of the expected coal production (approximately 100 million metric tons). Four more washeries are being installed, and additional capacity is currently under planning. Table 4.5 shows the details of coal washeries established in India up to 1970. In the Fifth Five-Year Plan, provision has been made for setting up 10 million metric tons of additional washery capacity, of which 4 million metric tons will be operational by 1978/79. A similar effort is necessary in coal beneficiation plants to improve the quality of high-ash Indian coal, both coking as well as noncoking, if the vast reserves of coal are to be utilized to their fullest extent.

All of the noncoking coal produced in India is used directly as fuel. This constitutes about 80 percent of the total, the balance being coking coal, which is used for production of coke of different grades and for various by-products, such as gas and tar. This means that less than 20 percent of the coal is carbonized to produce more valuable products. Coal-processing and conversion technologies would, therefore, require considerable expansion. Of the total coal tar production in the country, less than half is processed further for chemicals and other products, and the balance is still used as fuel.

Coal conversion technologies are not practiced in India at present. A beginning has been made by Coal India, by marketing smokeless coal for domestic consumption. Other potential technologies that require coal conversion include the following: (1) total gasification to produce fuel gas and synthetic gas; (2) briquetting and forming of coke; (3) low-degree carbonization, as for smokeless domestic fuels mentioned above; and (4) synthetic liquid fuels and coal tar chemicals. Two commercial plants for low-temperature carbonization are being installed in Andhra Pradesh and West Bengal. Two other total gasification plants for production of ammonia are being installed and are likely to start production in 1977. One farm coke plant (300 metric tons per day) on a demonstration scale is being installed at the Talcher coal field of Eastern India. The total expenditure on coal in the fifth plan has been planned at 1,025 crores of rupees as against earlier provision of 747.60 crores of rupees made in the draft Fifth Five-Year Plan. A large share of this expenditure will go into modernization of coal mines, for which a great deal of equipment is being manufactured indigenously, as well as being imported from Poland and other Eastern European countries.

In the early 1970s, coal constituted over 95 percent of the nation's commercial energy resources and contributed 60 percent of the total value of mineral production. This figure has changed recently due to upward revisions in the discovery of reserves of petroleum and increase in electric power-generating potential. Coal would now constitute between 85 and 90 percent of the total known commercial en-

TABLE 4.5

Coal Washeries, 1951-70

Location	Owner[c]	Starting Year	Capacity of Input (million metric tons/year)
West Bokaro	TISCO	1951	0.57
Jamadoba	TISCO	1952	1.44
Lodna	BCCL	1955	0.40
Kargali	NCDC	1958	2.72[a]
Durgapur	HSL	1960	1.50
Dugda I	HSL	1961	2.40
Bhowjudih	HSL	1962	2.00[b]
Patherdih	HSL	1964	2.00
Durgapur	DPL	1967	1.35
Dugda II	HSL	1968	2.40
Chasnala	IISCO	1968	2.00
Kathara	NCDC	1969	3.00
Gidi	NCDC	1970	2.84
Sawang	NCDC	1970	1.00
Total	—	—	25.62

[a] Expansion in 1966.
[b] Expansion in 1964.
[c] TISCO: Tata Iron and Steel Company; BCCL: Bharat Coking Coal Limited; NCDC: National Coal Development Corporation; HSL: Hindustan Steel Limited; DPL: Durgapur Projects Limited; IISCO: Indian Iron and Steel Company.

Source: Personally obtained from Central Fuel Research Institute.

ergy resources. About 80 percent of the production of coal in India is from underground mining; the average weighted seam thickness has been 4.1 meters and the average working depth approximately 138 meters. More than 75 percent of the reserves are said to be in seams over 4.8 meters thick. An indication of the production to reserves ratio in coal can be seen in the fact that in 1974, recoverable solid fuel reserves of India were 66 percent of those of Asia, excluding the Soviet Union, but production was only 2.4 percent of the world's total. Demand for coal, however, is sensitive to a large number of

ENERGY RESOURCES

variables, foremost among which is the price of coal and other energy sources. If the potential of coal is to be fully realized in the future, not only would large-scale investment have to be made in coal-mining equipment, beneficiation plants, and other processes for making coal more easily "usable," but pricing and other policy options would have to be directed to stimulate a greater increase in demand.

Some of the problems of the coal industry, particularly its lack of adequate modernization, are related to the history of this industry. The mining of coal in India is supposed to have started as far back as 1774, when simple, shallow mines were developed in the Ranigunj coal field in Eastern India. Systematic quarrying appears to have started only in the second quarter of the nineteenth century. In the second half of the nineteenth century, approximately 50 collieries were in operation, producing about 2.8 lakh metric tons of coal per annum in the Ranigunj area. This production level had grown almost ten times toward the end of the nineteenth century. In the meantime, mining was also undertaken in other areas of East India and Central India. By the year 1900, coal production had grown to a total of 6 million metric tons. The growth of production in this industry is shown in Table 4.6.

TABLE 4.6

Growth of Coal Production, 1900 to 1973/74
(millions of metric tons)

Year	Coal Production
1900*	6.00
1920	17.80
1930	23.80
1940	29.39
1950	32.81
1955	38.84
1960/61	55.70
1965/66	67.66
1969/70	75.74
1970/71	72.96
1973/74	77.90

*Production statistics were earlier maintained by calendar years.

Source: Report of the Fuel Policy Committee (New Delhi: Government of India, Controller of Publications, 1974), Table 7.4, p. 44.

TABLE 4.7

Distribution of Coal Mines, by Size, 1970/71

Size/Production per Year[a] (metric tons)	Number	Annual Total Production (millions of metric tons)
Up to 6,000	228	Less than $\frac{1}{2}$[b]
Between 6,000 and 12,000	47	Less than $\frac{1}{2}$[b]
Between 12,000 and 60,000	211	7
Between 60,000 and 120,000	100	9
Between 120,000 and 300,000	130	25
Between 300,000 and 600,000	74	28
Above 600,000	7	5

[a] As per the licensed or approved production capacity.
[b] Exact production data not available for small coal mines included in this category.

Source: Report of the Fuel Policy Committee (New Delhi: Government of India, Controller of Publications, 1974), Table 7.5, p. 44.

The neglect of the industry was as much responsible for its lack of modernization as for the existence of a very large number of small mines. The FPC carried out a survey of mines in 1971 and found that the total number in operation was over 800 (distribution by size is shown in Table 4.7). From this, it is apparent that the 480 mines combined together had a total production of only 8 million metric tons. In the early 1970s, after reorganization and nationalization, large investments were made to modernize the equipment and to mechanize the operations. The total amount of people employed in the coal-mining industry is around 0.35 million, with very low levels of productivity. The output per mine shift increased from 0.35 metric tons in 1951 to 0.48 metric tons in 1961 and to 0.67 metric tons in 1971. The current level of productivity is around 0.8 metric tons per mine shift.

The coal industry in the country is going through a major change at present and is gradually becoming more dynamic, mainly as a result of superior management and policy making (at its highest levels). Improvements are essential in order to ensure that supplies, which have generally been ahead of demand in the past, will continue to be so in the future. In the opinion of various technical experts, the present level of mechanization and facilities in the coal industry will not permit the annual production level to exceed 312 million metric tons by 1990/91. This may well turn out to be adequate for meeting demand in that year, unless a major shift in consumption in favor of coal takes place. The possible levels of production from different coalfields according to expert opinion are shown in Table 4.8. Even though the demand levels anticipated in the report of the FPC[6] and the study by Kirit Parikh[7] are not likely to materialize, it is not inconceivable that with large-scale efforts, the government could bring about a major change in consumption patterns. For instance, the natural hesitation that consumers have in using coal for domestic consumption can be largely overcome by gasification of coal and supplying coal gas through a network of pipelines. This, however, is not yet commercially feasible and will be discussed further in the next chapter. With the greater expenditure planned for power generation through thermal plants, increases in electricity consumption will also be accompanied by increases in coal consumption.

HYDROELECTRIC RESOURCES

India is well endowed with geographical features and river systems that can enable it to generate a large amount of power from hydroelectric sources. Soon after independence, the Central Water and Power Council carried out a series of surveys on the basis of which the hydroelectric resources of the country were mapped and estimated. A total of 260 specific schemes were evaluated, with groupings based on river systems, as follows:

1. The western-flowing rivers of southern India, covering the Western Ghats and coastal strip;
2. The eastern-flowing rivers of southern India, going through the states of Andhra, Karnataka, and Tamil Nadu;
3. The rivers of Central India, which flow both eastward and westward;
4. The Ganges Basin, including all its tributaries in the Himalayan regions and the plains;
5. The Bramhaputra Basin and neighboring areas, extending over Assam, Meghalaya, and West Bengal; and
6. The Indus Basin, covering tributaries of the Indus River in the northwestern part of India.

The exploitable hydroelectric power potential for each of these river systems was estimated and a geographical distribution arrived

TABLE 4.8

Possible Levels of Production of Coal from Different Coalfields
up to 1990/91
(millions of metric tons)

Coalfield	1978/79	1983/84	1990/91
Makum/Northeast region	1	2	3
Bengal-Bihar			
Jhari	26	36	50
Mugma and Ranigunj	33	45	60
East Bokaro	7	12	13
South Karanpura, West Bokaro, and Ramgarh	12	24^a	30^b
North Karanpura (including Hutar and Daltongunj)	—	15	22
Rajmahal	—	3	10
Singrauli (Madhya Pradesh and Uttar Pradesh	7	15	24
Talcher (Orissa)	4	8	12
Sohagpur and Korea-Rewa	15	19	20
Korba-Hasdeo and Ib River	5	12	22
Pench-Kanhan-Tawa	7	7	8
Kamtee (Maharashtra)	2	5	7
Chanda-Wardha	4	6	10
Godavary Valley	12	16	21
Total	135	225	312

^{a}Coking, five.
^{b}Coking, seven.

Source: Report of the Fuel Policy Committee (New Delhi: Controller of Publications, 1974), Table 7.7, p. 46.

at to indicate the power potential from hydroelectric sources for each state. These are shown in Table 4.9. The potential shown in this table has been calculated on the basis of a 60 percent load factor. In actual fact, establishing a larger number of hydroelectric power stations would result in these being utilized only for peak purposes. This would lead to a much lower load factor being achieved for hydroelectric power. The energy potential of these would, therefore, vary according to the extent and nature of utilization. The assumption of 60 percent load factor would lead to regional generation of power of the levels indicated in Table 4.10. One factor that generally weighs against large-scale expansion of hydroelectric power potential is the requirement of large investments in such schemes. Fortunately, in the case of India, approximately 25 percent of the total potential indicated in Table 4.9 is of the "run-of-the-river"* type. This potential could very well increase with development of new technologies in hydroelectric power potential. For instance, the commercial feasibility of low heads of water for power generation would result in a large number of potential sites becoming available. Projects utilizing traps

TABLE 4.9

Distribution of Power Potential, by State

State	Megawatts at 60 Percent Load Factor
Andhra Pradesh	2,476.5
Assam (including Meghalaya, Nagaland, and Mizoram)	11,599.4
Bihar	609.7
Gujarat	677.0
Jammu and Kashmir	3,590.5
Kerala	1,539.5
Madhya Pradesh	4,582.3
Madras	708.2
Maharashtra	1,909.6
Mysore	3,372.8
Orissa	2,062.0
Punjab and Haryana	1,360.5
Rajasthan	149.0
Uttar Pradesh	3,764.0
West Bengal	22.0
Himachal Pradesh	1,867.5
Manipur	865.0
Total	41,155.5

Source: Report of the Fuel Policy Committee (New Delhi: Controller of Publications, 1974), Table 4.5, p. 27.

*Small capacity plants that have no upstream storage of water are usually called run-of-the-river plants. These can be established with relatively low investment levels and short gestation periods.

TABLE 4.10

Hydel Energy Potentials of Various Regions

Region	Billions of Kilowatt Hours	Percentage of Total
Eastern	14.2	6.5
Northern	36.6	16.9
Central	43.9	20.3
Western	13.6	6.3
Southern	42.6	19.7
Assam	65.5	30.3
Total	216.4	100.0

Source: Report of the Fuel Policy Committee (New Delhi: Controller of Publications, 1974), Table 4.6, p. 27.

from 30 to 300 meters account for 23.86 million kilowatts and low-head projects, ranging from 8 to 30 meters, have a total potential of 3.66 million kilowatts. The dispersal of hydroelectric potential all over the country enables accessibility of all regions to some hydroelectric potential site or the other. A detailed mapping of present hydroelectric potential in the country is called for. Particularly in the Himalayan region, there is a great likelihood of upward revision of the estimated hydroelectric potential. Furthermore, most of the sites that have been included in the above estimates are those that have a year-round potential for power generation. In addition, there exist all over the country a large number of sites that could provide power on a seasonal basis. If these were to be considered, the seasonal potential of hydroelectric power would be revised upward by a fairly large amount. The Power Economy Committee, established by the Indian government, expressed the view that "on the basis of the latest information regarding hydel energy resources and their economics of development, it would be possible to install about 80 to 100 million kilowatts of hydel capacity on our river systems during the next two or three decades."[8]

NUCLEAR RESOURCES

India has emerged as one of the major countries of the world in nuclear technology and application. The story of atomic energy in India is a very interesting case of how programs are generally built around the personalities of individuals. India was fortunate in having a brilliant and resourceful scientist in Homi Bhabha, who was given charge of the Atomic Energy Commission's DAE by India's first prime minister, Pandit Jawaharlal Nehru. Bhabha used his vision and rapport with the prime minister in forging an effort that was unprecedented for a developing nation that had only recently broken the shackles of colonial rule.

Seen in the context of the First Five-Year Plan, there did not appear to be any great economic argument in favor of developing nuclear power. India's electricity production was around 2.3 million kilowatts (against the present level of over 20 million kilowatts). The oil resources existing at that time were meager, but there did not appear to be any great anxiety on the part of the government in the early 1950s with regard to likely shortages in the supply of oil.

Critics of the nuclear program in the early 1950s saw in Bhabha's efforts only an attempt to build an empire and buy the luxury of a nuclear program at the cost of other priority areas. Bhabha's arguments, which were supported with strong backing by the prime minister, finally made headway. He made a case in favor of nuclear power on the basis that even though India had substantial reserves of coal, its poor quality would rule out coal as a main source of energy for the future. In Bhabha's opinion, the gap between the aggregate of oil, hydroelectric, and coal resources had to be made good by nuclear power.

This was also related to the fact that coal deposits in India occur mainly in the eastern region and, therefore, for energy supplies, the western regions would have to depend on other sources. If a circle with a radius of 500 kilometers is drawn with the heart of coalfields in eastern India as the center, it will be found that 35 percent by area and 30 percent of the country by population lies outside the circle. Significantly, the states that lie outside the circle are those where agriculture and industry are vital to the development of the country (Maharashtra, Gujarat, Punjab, Kerala, and Tamil Nadu).

With the high-ash content of Indian coal and the distances required to feed these regions, nuclear energy acquired an edge over other forms for some of these regions. It was also felt that in order to increase consumption of coal, a great deal of indirect expenditure would have to be incurred, such as in modernization of the railway system, which is basically capital intensive. Before the advent of diesel and electric traction on the railways, 30 percent of the coal carried by the railways was used for their own consumption.

Undoubtedly, India's decision to develop a strong nuclear program rested as much on considerations of national esteem and power politics as on its long-run economic benefits. Nuclear power plants possess substantial economies of scale, which could be taken advantage of in supplying power to those regions where demand was concentrated in a relatively small area. Various case studies and feasibility reports were prepared to show the economic benefits of nuclear power. One detailed case study pertains to a large agro-industrial complex in western Uttar Pradesh. This envisaged the production of nuclear power to be used subsequently to produce fertilizers, on the one hand, and to operate tube wells and feed agro-industries with power, on the other. The scale of the plant visualized was 1,200 megawatts of electric power (mwe),* and the consumption pattern that was visualized is indicated as follows:

Operation	Mwe
Fertilizers	775
Tube wells	300
Aluminum processing plants	125

Some of the relevant estimates are summarized in Tables 4.11-4.13. With these costs and by evaluating the benefits, it was shown that the returns from investment in nuclear power promised to be as high as 30 percent.

Bhabha died tragically in a plane crash in 1965. He was succeeded by Vikram Sarabhai, another eminent scientist and industrialist, who maintained the momentum that Bhabha had generated in India's nuclear program. (Skeptics of the nuclear program were often silenced by the great authority that the chairman of the DAE was given by the government.) It is likely that even if the economic benefits, as demonstrated by the table above, were not apparent, the nuclear program would have been pushed along vigorously, as it was in the 1950s and 1960s. The strategy conceived by Bhabha and continued subsequently was fairly simple. He realized early that India's uranium reserves were meager as compared to the rest of the world, but on the other hand, deposits of thorium were substantial (in fact, India is one of the largest repositories of thorium). It was, therefore, established that nuclear power strategy in India should aim at surveying for uranium, while at the same time, moving toward greater consumption of thorium.

This led Bhabha to work out a three-stage program. The first involved a chain of nuclear power stations of the Canadian design, popularly known as "Candu type." Candu reactors use natural uranium as fuel and heavy water as moderator. This type of reactor serves two purposes—it produces electric power and, at the same time, con-

---

*Mwe is used to indicate megawatts of electric power as opposed to mw(th) which indicates megawatts of thermal power.

TABLE 4.11

Plant Investment Costs in the Agro-Industrial Complex
(in 1976 prices)

Plant	Capacity	Cost (crores of rupees)*
Power station	1,100 mwe	223.90
Fertilizer plants	1.46×10^6 mt/annum	131.60
Aluminum plant	5×10^4 mt/annum	54.55
Total for industrial complex		410.05

*One crore equals 10 million.

Source: Personally obtained from officials of the Department of Atomic Energy.

verts a part of uranium 238 in natural uranium into fissile material plutonium 239. In balance, even the amount of plutonium so produced is far less than the uranium 235 consumed. With a large number of plutonium-producing stations, enough stocks could be generated to enter into the second stage of Bhabha's program.

This second stage calls for the fast breeder reactor (FBR). The FBR was envisioned as serving two purposes, that is, for production of power and, at the same time, for conversion of metal into fissionable material. The FBR technology that is being developed in

TABLE 4.12

Details of the Agricultural Side of the Agro-Industrial Complex

Factor	Amount
Area proposed to be irrigated	1.5×10^6 ha.
Number of shallow-tube wells	13,000
Number of deep-tube wells	13,000
Annual income per hectare	Rs 8,275
Net annual profit per hectare	Rs 4,810

Source: Personally obtained from officials of the Department of Atomic Energy.

TABLE 4.13

Overall Capital Costs of the Agro-Industrial Complex[a], 1976

Item	Cost (crores of rupees)[b]
Power plant	237.1
Fertilizer plants	134.0
Aluminum plant	54.6
Rural electrification and tube wells	220.3
Transport facility improvements	113.7
Rural credit and so forth	133.3
Agricultural implements and warehouses	339.5
Total	1,232.5

[a] Includes working capital.
[b] One crore equals 10 million.

Source: Personally obtained from officials of the Department of Atomic Energy.

India was expected to dovetail into the first stage of nuclear development. It was intended that the FBR would be employed mainly to convert the uranium 238 left over by the Candu-type reactors into uranium 239. This would later be used for the thorium to uranium 233 conversion also. In the third stage, technology was expected to be developed to change over completely to a thorium-uranium cycle. The elements of Bhabha's strategy are shown in the flow chart in Figure 4.3. The new program, which rested on the strategy mentioned above, was charted by India in its various elements, and a large number of projects and schemes were launched toward the overall objective outlined above.

The first achievements in the nuclear program were the establishment of the Bhabha Atomic Research Centre (BARC). This center deals with basic research and development work being done by the DAE. The activities undertaken in this establishment have included: (1) acquisition of skills in reactor physics and reactor technology, (2) evolving techniques for production of nuclear-grade reactor materials, (3) evolving techniques for cladding, which is a process of coating one metal (in this case, nuclear fuel rods) with another such as aluminum or zirconium to prevent escape of fission products and oxidation of the fuel, and fabrication of fuel elements, (4) evolving techniques for reprocessing of irradiated fuels, (5) evolving methods

FIGURE 4.3

Strategy for Nuclear Power in India

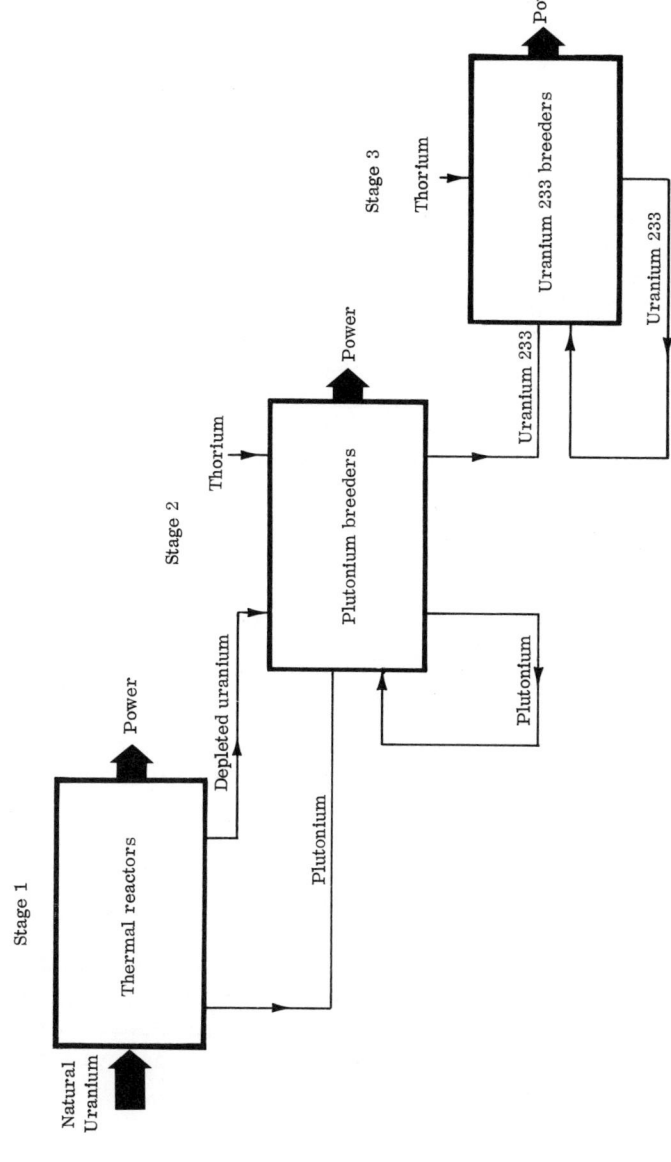

Source: Compiled by the author.

for radioactive waste management, and (6) developing procedures and techniques for protection of human beings from nuclear hazards.

An extension of these facilities was set up in the form of the Nuclear Fuel Complex at Hyderabad, a manufacturing facility for instrumentation in reactor control in the form of the Electronics Corporation of India, Hyderabad, and a processing facility for spent fuel from the Tarapur plant. In addition, a large number of skilled scientists, engineers, and technicians have been trained at BARC and have taken up various positions in the different wings of the DAE.

The nuclear power program in the country commenced with the establishment of the Tarapur Atomic Power Project (TAPP). This project deviated from the strategy outlined above, since it did not buy a reactor of the Candu type. It uses the nuclear energy in uranium 235 as fuel and light water as moderator, in contrast with natural uranium and heavy water in the Candu-type reactor. TAPP was executed by General Electric of the United States as a complete turnkey* project. But a large number of components for the plant were subcontracted to Indian industry and the DAE, which promoted development of skills and know-how. The great benefit of TAPP was the experience in nuclear power plant construction, installation, and operation that Indian scientists and engineers were able to gain. TAPP has been the subject of a great deal of controversy. Its functioning, in keeping with first-generation plants around the world, has been far from satisfactory and has led to extensive criticism. Although there have been extended shutdowns in this plant, the power produced from it has been supplied at prices competitive with fossil fuel power stations in the states of Gujarat and Maharashtra.

At the same time that the Tarapur station was being established, the DAE launched its Candu program. The first phase of the Rajasthan Atomic Power Project (RAPP) has already been completed. The performance of this plant thus far has been highly satisfactory. This particular plant shows how the establishment of sophisticated technologies cannot take place in isolation from the industrial environment in a developing country. Since the attempt of the DAE in this case was to promote a large indigenous content in the plant, various obstacles were faced in trying to convince leading manufacturers of various equipment in India to contract for their share in this effort. Delays in negotiations and contractual formalities lasted for a number of years. Problems of quality control also resulted in considerable delays. Delays also took place due to transportation bottlenecks. A

---

*A turnkey project is one that is completely designed and executed by the contractor and handed over to the buyer for operation after completion.

large number of parts required for the plant were extremely heavy, and movement facilities did not exist to transport them from Bhopal, the place of manufacture in Central India, to Kota, in Rajasthan. As an example, the two reactor end-shields each weighed about 120 metric tons. These had to be moved by a special train, with hydraulic jacks and lifting devices to enable the shields to clear the surface and overhead obstructions along the railway track. In some cases, even, the railway track had to be relaid, and the structures on which the tracks were laid had to be strengthened. The second reactor of RAPP has also recently been completed, and the DAE's efforts are now concentrated on the Madras Atomic Power Project (MAPP), which will have two reactors of 230 mwe capacity each. It is expected that MAPP will be operational toward the end of the Fifth Five-Year Plan, that is, in 1978/79.

At the same time as the development of MAPP is taking place, work also has been started on another Candu-type station at Narora, in the northern state of Uttar Pradesh. Narora would function basically as a power station only, and it is not likely to be used as the nucleus of an agro-industrial complex. At present, work has not proceeded beyond the Narora station. A scaling down of targets for nuclear power generation has taken place in the light of the experience of the past four or five years, mainly because of the inadequate support from the supporting industries, which are required to provide inputs for nuclear power station construction. As a result, nuclear power potential in the country by around 1980 is likely to be not more than 1,600 mwe. This compares unfavorably with the grandiose figure of 8,000 mwe that was suggested as the target for 1980 by the Atomic Energy Commission as far back as 1954. Subsequently, the Energy Survey Committee in 1965 forecast a figure of 5,000 mwe by 1980, and in fact, the figure of 1,600 mwe, which is likely to be realized, is far below the capacity of 2,700 mwe that was suggested by the Atomic Energy Commission in the late 1960s. The biggest constraint in rapid development of nuclear power potential is the lack of availability of adequate resources. Even though the capacity of the capital industries involved in supplying equipment for nuclear power stations has not been up to original expectations, there is some satisfaction in observing that, technologically, India has progressed remarkably well in this complex field.

Serious effort is being made in the development of the breeder reactor. The technological problems in establishing breeder reactor technology on a commercial scale are vast and complex. This is an area in which even the more advanced countries are still far from being in a satisfactory stage of development; there are at present not more than ten breeder power stations in operation all over the world. To provide a focus for the development of breeder technology, the

DAE started the Reactor Research Centre at Kalpakkam in Madras. This Reactor Research Centre is likely to play the same role in developing indigenous breeder technology as BARC played earlier in the development of nuclear reactors. The first stage of this research center's activity will be the establishment of a fast breeder test reactor (FBTR), which is presently under construction. Using this reactor as a model, a large number of experiments and tests will be carried out, particularly for developing fuel for breeders, reprocessing fuel irradiated in breeders, and in developing various types of materials required in the structure of a breeder reactor.

This type of research and development is an essential prerequisite in breeder technology, since components of this reactor are subject to the possibility of great damage due to the large number of neutrons that are present constantly. This damage happens on a very much larger scale than in the case of Candu-type reactors. The very high power densities in a FBR, of the order of 500 mwe per cubic meter of core volume, call for cooling substances, like liquid metals. The use of liquid sodium for this purpose has been accepted by scientists all over the world. Attached to the use of sodium itself, for this purpose, is the development required in manufacture of sodium pumps, sodium to sodium heat exchangers, sodium to water steam generators, and a variety of sophisticated instruments to monitor the sodium systems. Since the plutonium-bearing fuel elements are very expensive, they must have high fuel ratings and long fuel life, in the neighborhood of 50 to 200 mwe per metric ton rating.

The progress of technological developments in the area of fast breeder technology in India has been reasonably satisfactory. In 1972, a preliminary pulsed fast reactor became "critical" at BARC. According to present indications, the low-power FBTR at Kalpakkam is likely to be commissioned by 1980 or so. The DAE plans on completing its first 500 mwe commercial breeder reactor by 1986, and thenceforth, their growth is expected to be very rapid. Some scientists in the field of atomic power in the country visualize a number of 1,000 mwe breeders in operation, with a total installed capacity of perhaps 10,000 mwe by the turn of the century.

The sum total of the Indian experience in nuclear power developments indicates that even if technological competence can be generated within the country, there are various constraints, in the form of inadequate resources and the slow growth of equipment-manufacturing capacity. There is also a great deal of uncertainty about the long-run economic benefits of nuclear energy. Proponents of nuclear power come up with estimates that indicate that this form of energy is almost half as expensive as power generated from other forms, whereas those who are against rapid expansion of nuclear power capacity contend that for a poor country such as India, a capital-intensive technology, such as nuclear power generation, does not take into account the opportunity cost of capital. In their view, an emphasis should be placed on other forms of energy generation, such as biogas plants, solar energy, and so forth. Political realities and the prestige attached to nuclear power generation will, however, play a significant part in government policy making, and with the other unconventional forms of energy still in the beginning stages of development, an expansion of nuclear power resources in the country is bound to take place in coming decades.

SOLAR ENERGY RESOURCES

By virtue of its being in the tropics, India receives a large amount of solar energy throughout most of the year. The potential available from this power source can be gauged from the fact that one square meter of area receives about one kilowatt of solar power for 2,500 hours per year on an average. In terms of oil equivalent, five square meters of area receives as much energy in a year as can be obtained from one metric ton of oil. India's total energy needs are, therefore, available and can be satisfied by exploiting the solar energy received on an area of 4,000 square kilometers at 20 percent conversion efficiency. This represents 0.15 percent of the total area of the country. These figures, of course, can at best represent the ideal, which is nowhere near achievement, given the current state of technology. Whereas solar energy represents the final inexhaustible form of energy on this planet (as long as sunshine exists), the world is nowhere near finding solutions that would be economically competitive in the use of solar energy as against other known forms of energy.

Solar energy has been used to a small degree in a large number of applications throughout history. A common application, which is evident even today, is the evaporation of water in the salt industry for production of salt. Other applications are water heating; desalinization of water by distillation; drying of various products, such as textiles; and the use of solar cells. The potential, however, for a variety of applications is considerable. For almost 20 years now, Japan and Israel have been using solar energy for producing hot water. Even in the United States, experiments have been carried out, and a number of homes are now (fashionably) using solar heating for their domestic requirements. The various technologies that could make solar energy relevant to our life-styles will be discussed in the next chapter. At this stage, it will be sufficient merely to mention some of the work that has already been done in the country to provide an indication of the reality of solar energy as a resource that can be harnessed for productive activities. The potential that is available can be converted into useful applications only through a widespread program of research and development. Various institutes have done work in this field, and their achievements are listed below:

1. The Central Salt and Marine Research Institute at Bhavanagar has done extensive work on investigating solar radiation, solar distillation, and so forth, and results achieved by them have been applied in the production of salt in different parts of the country.
2. The Central Building Research Institute at Roorkee has evolved a design of a flat plate solar energy collector and solar water heaters, which even though not produced on a large commercial scale at present, are likely to be prescribed for installation in homes in the colder parts of the country.
3. The Defence Research Laboratory at Jodhpur has done a considerable amount of work in using solar energy for various residential applications that are useful to living at high altitudes.
4. The Solid State Physics Laboratory, under the Council of Scientific and Industrial Research, is doing pioneering work in the design of various solar cells that can be used for industrial and residential applications.

In recent years, a number of academic institutions and industrial organizations have taken up work in the application of solar energy for residential and industrial purposes. The Indian Institute of Technology at Madras and BHEL have been working jointly on the development of a solar energy electric generator. Other institutions that have been working in this field are Jadavpur University, near Calcutta, the Central Arid Zone Research Institute, Jodhpur, the Indian Institute of Science at Bangalore, and various regional engineering colleges and institutes of technology.

The applications of solar energy for useful purposes would require funding on an extensive scale in order to make this source a reality in the next few decades. This is an area in which the sharing of information and research results would be of vital interest to a developing country like India. Intensive research is now being undertaken in the field of solar energy application in the United States, Australia, Israel, and other countries; India could derive substantial benefits by using technologies evolved in these countries in the next few years.

BIOGAS RESOURCES

The source of energy that is being advocated with considerable vigor these days as being the biggest and most important answer to

Indian energy problems is the use of biogas, or gobar gas, plants.*
India has had a considerable lead in the technology of gobar gas plants.
On the basis of estimates available to the Indian government, approximately 170 million metric tons of dry cattle dung is produced annually in the country today. Of this, approximately, one-third is used as fuel in most rural, as well as some urban, households. (In the household, using cattle dung directly for cooking purposes presents an inefficient form of energy conversion.)

The technology of gobar gas plants raises the efficiency levels of energy conversion and also produces manure, which can substitute for fertilizers produced by energy-intensive industrial processes. The pioneer of gobar gas plants has been the Khadi and Village Industries Commission of India (KVIC). This commission, apart from extending technical assistance and marketing gobar gas plants of various sizes has also been granting loans and other forms of financial assistance for the construction and setting up of such plants. The technology as it exists today, however, is not free from controversy about its technoeconomic merits.

Initial estimates indicated that gobar gas plants were justified for use in all states in India, except for the states of Assam, Jammu and Kashmir, Madhya Pradesh, and Orissa, which have relatively richer forest resources. Research activities of various technical bodies, such as the National Committee for Science and Technology, have led to improvements in existing designs of gobar gas plants; these improved designs have led some reductions in cost of manufacture and installation.

A survey of gobar gas plants installed in the country indicated that in spite of efforts of the KVIC, only 6,158 plants had been installed in the country by 1973/74. Of these, a very large percentage were either not being used at all or were working at considerably less than full capacity. As such, gobar gas plants have not really made a serious dent in the rural energy problem of the country. The huge potential that exists as a result of the large amount of dung that is produced in the country cannot be tapped merely by setting up a large number of gobar gas plants; the infrastructure that would ensure proper utilization of existing plants must be provided. In the last four years, various banks and other financial institutions have been offering loans at lower interest rates to consumers setting up gobar gas plants in the country. As a result, therefore, by March 31, 1975, the total number of plants in the country had reached 12,139. The performance of various states in this regard has, however, not been uni-

---

*Dung-based biogas plants are referred to popularly as gobar gas plants.

TABLE 4.14

Distribution of Biogas Plants, by States, as of March 31, 1975

State	Number of Plants	Percentage of Total
Andhra Pradesh	816	6.72
Assam	23	0.19
Bihar	143	1.18
Gujarat	3,416	28.14
Haryana	2,181	17.97
Himachal Pradesh	7	0.06
Karnataka	1,054	8.68
Kerala	306	2.52
Madhya Pradesh	417	3.44
Maharashtra	2,018	16.62
Orissa	51	0.42
Punjab	198	1.63
Rajasthan	111	0.91
Tamil Nadu	520	4.28
Uttar Pradesh	805	6.63
West Bengal	51	0.42
Delhi	6	0.05
Goa	7	0.06
Pondicherry	6	0.05
Jammu and Kashmir	2	0.02
Tripura	1	0.01
Total	12,139	100.00

Source: The National Committee on Science and Technology, Ministry of Science and Technology, unpublished.

form, as Table 4.15 indicates, in which the distribution by state of existing gobar gas plants on March 31, 1975, is shown.

The potential presented by this source of energy could be realized by a three-pronged program directed mainly at rural areas. This would include organization of technological support, maintenance and operational support, and extension and training support. In order to organize these services, widespread studies would have to be undertaken to find out the problems being faced by farmers and owners of gobar gas plants all over the country. Such studies are now being undertaken by a number of institutions. A very useful piece of work

ENERGY RESOURCES 103

on the subject has been done by the Centre for Management and Agriculture at the Indian Institute of Management at Ahmedabad.[9] This group concluded that the missing link in making gobar gas plants work was the organizational aspect, for which very few agencies in the country were either capable or willing to tackle.

The abundance of dung produced in the country and India's preeminence in the technology and practice of running gobar gas plants makes this energy source an attractive alternative. In 1976, Senator Gaylord Nelson of the United States introduced a bill to authorize a study by the U.S. Department of Agriculture on anaerobic digestion technology for possible application in the United States. In the proceedings related to the introduction of this bill, the example of India was mentioned as a pioneer in the conversion of cow dung to useful energy (it was also mentioned that a large number of anaerobic converters were being successfully used in India by 1976). It would indeed be unfortunate if the vast potential available to the country in this field is not fully tapped because of lack of organizational and administrative effort.

WIND POWER RESOURCES

A cursory look at India's coastline would indicate a large potential for wind power and its application. Historically, this form of power has been known since very early times, when ships sailed all over the oceans, using the power of wind. Unfortunately, until fairly recently, very little development took place in this country in the use and practice of wind power technologies. In fact, even though other countries in the world have been using windmills for centuries, this technology has been almost nonexistent in India. The answer may perhaps lie in the nature of wind currents and movements in the country. Surveys of wind patterns in the country reveal that winds on this subcontinent are largely seasonal. The predominance of wind currents seem to occur during premonsoon and postmonsoon periods. In addition, the average speed of Indian winds is in the region of 10 to 20 kilometers per hour, and this is lower than that obtaining in such countries as Canada, Holland, Iran, and China, which have had a tradition of using windmills. Due to the seasonal nature of wind currents, the capacity utilization of windmills would perhaps be no greater than 25 percent. (This, of course, could change with improved designs of windmills, since existing designs are generally made for running at wind speeds higher than those obtained in India.) With lower wind speeds, in order to produce substantial outputs of power, Indian windmills would have to be of very large and bulky designs. This, of course, would increase the investments required for using wind power.

TABLE 4.15

Distribution of the Forest Area, by State, 1969/70

State/Union Territory	Forest Area (millions of hectares)	Percentage of Forest Area to Geographical Area	Exploitable Area[a]	Potentially Exploitable Area
State Territory				
Andhra Pradesh	6,512	22.53	4,836	1,676
Assam (including Meghalaya and Miziram)	4,442	36.38	1,509	200
Bihar	3,059[b]	17.59	2,107	952
Gujarat	1,800	9.18	1,513	—
Haryana	142	3.21	57	30
Himachal Pradesh	2,159	38.78	1,203	447
Jammu and Kashmir	2,105	9.47	1,858	246
Kerala	1,269	32.66	108	224
Madhya Pradesh	16,813	37.97	10,122	3,987
Maharashtra	6,696	21.76	3,829	2,471[c]
Mysore	3,510	18.30	2,615	895
Nagaland	290	17.54	31	207
Orissa	6,746[d]	43.29	5,383	1,042
Punjab	202	4.01	90	—
Rajasthan	3,760[d]	10.99	2,810	—
Tamil Nadu	210	16.99	1,419	791
Uttar Pradesh	4,872[d]	16.55	3,462	701
West Bengal	1,183	13.47	1,080	65
Manipur	602	26.92	291	311
Tripura	630	60.11	240	190
Union Territory				
Andaman and Nicobar Islands	747	90.11	503	243
Dadra and Nagar Haveli	21	42.86	21	—
Delhi	5	3.36	3	—
Goa, Daman, and Diu	105	27.56	95	10
Arunachal Pradesh	5,154	61.67	34	52
Other	—	—	—	—
Total	75,033	22.87	45,579	14,760

[a] Forest in use.
[b] 1967/68 figures repeated.
[c] Estimated.
[d] 1968/69 figures repeated.

Source: Task Force on Forest Development, Report of the Fuel Policy Committee (New Delhi: Controller of Publications, 1974), Table 4.9, p. 32.

Therefore, wind power in India does not present a major potential source of energy conversion. Except for isolated areas, such as in Rajasthan and some coastal areas of South India, wind power does not appear to have a very great future, given the present state of the relevant technology.

FOREST RESOURCES

Even though, from a social point of view, the use of forest energy is very expensive, the lack of distribution systems and nonavailability of commercial sources of energy, particularly in rural areas, warrants the consumption of forest resources as fuel (this is an imperative for a number of years to come). Table 4.16 shows the breakup of forest area in the country by states in 1969/70. Efforts at forestry development, and in particular the planting of fast-growing forests, can lead to an increase in this area, if accompanied by determined execution of a conservation policy. The gravity of the problem can be seen from the fact that about 10 percent of the fuel wood used in the country comes from authorized production of timber and firewood. Almost 90 percent of the total of over 1 million metric tons comes from unauthorized and unrecorded fellings, both from forest as well as nonforest areas. If urgent steps are not taken to regulate the consumption of firewood in the country, there will be disastrous consequences in the next few decades. This can only be prevented by a determined program of afforestation, using quick-growing trees and better policing of fellings all over the country, as well as by opening up rural markets to alternative cheap forms of energy.

This chapter has attempted to provide an insight into available resources for energy production in this country. These resources can be classified into priority areas, where their exploitation is within reasonable reach of society given its present state of development, and those where substantial investments would have to be made in technologies, infrastructure, and organizational inputs. Naturally, a national energy policy for India would have to address itself to problems of short-term energy supply, as well as of a long-term nature, which would deal with the use of these resources by investments in research and development and other developments of the type mentioned above. The following chapter will go into the technological aspects of energy production and consumption, as well as some of the technological alternatives available to India.

NOTES

1. Energy Survey of India Committee, <u>Report</u> (New Delhi: Government of India Press, 1965).

2. National Council of Applied Economic Research, Demand for Energy in Eastern India (New Delhi: National Council of Applied Economic Research Press, 1963); National Council of Applied Economic Research, Demand for Energy in Western India (New Delhi: National Council of Applied Economic Research Press, 1965); National Council of Applied Economic Research, Demand for Energy in India (New Delhi: National Council of Applied Economic Research Press, 1966); National Council of Applied Economic Research, Domestic Fuels in India (London: Asia Publishing House, 1959).

3. Fuel Policy Committee, Report (New Delhi: Controller of Publications, 1974).

4. Michael Tanzer, The Political Economy of International Oil and the Underdeveloped Countries (London: Maurice Temple Smith, 1969), p. 238.

5. Fuel Policy Committee, Report, p. 23.

6. Ibid., p. 45.

7. Kirit Parikh, Second India Studies: Energy (New Delhi: Macmillan Company of India, 1976), p. 50.

8. Fuel Policy Committee, Report, p. 28.

9. T. K. Moulik and U. K. Srivastava, Bio-gas Plants at the Village Level: Problems and Prospects in Gujarat (Ahmedabad: Indian Institute of Management, 1975), pp. 124-26.

CHAPTER 5

TECHNOLOGICAL ALTERNATIVES

The energy crisis brought home the importance of developing technologies for the production and application of energy. An awareness and assessment of future demand for energy and the resources available for meeting it clearly indicate that new technologies are the only answer to the world's energy problems. In this chapter, we consider some of the important areas in which technological developments are taking place, and an attempt is made to assess the broad implications of various technological alternatives facing India and the rest of the world. It must be mentioned that the global nature of the energy problem makes technologies in developed countries of equal importance to developing countries as well. In this chapter, therefore, although we have concerned ourselves mainly with developments in India, examples from other countries have been presented.

There are certain aspects of technological improvements that should be considered in dealing with conventional forms of energy. These are important but generally well known and have been researched by different authors. Of greater relevance to the solutions of problems in the future is the question of new forms of energy and their applications. We have placed greater emphasis on new technologies and alternatives that India faces or is likely to face in the future and for which research and development efforts are either being made or must be made in the years to come. Mention wherever necessary of conventional technologies and their applications has been kept brief.

TECHNOLOGICAL ALTERNATIVES TO ENERGY FROM COAL

The abundant reserves of coal available in India and the production that exists for producing this commodity makes it a natural choice as the major source of commercial energy in the country. The increase in production and consumption of coal planned for by the government can, however, take place only through sustained efforts in improving technologies related to coal consumption and production. Long-term strategies and plans have been evolved by the Ministry of Energy and Coal India; they will concentrate on research and development efforts in improving technologies for exploration, mining, equipment engineering, beneficiation, and coal usage. The formulation and implementation of energy policies with respect to coal become more manageable as a result of the nationalization of coal mines in the country. At the same time, it places a tremendous responsibility on the government of India for drawing up appropriate policies and taking actions that the market may not be able to support.

The first challenge facing the coal industry in the country is upgrading the technology of the production of coal itself. Efforts have to be made not only to introduce the method of long wall mining but to mechanize existing bord and pillar methods also. It is also intended to take advantage of economies of scale by planning for mines to grow larger by combining two or more mines into one wherever necessary. Attendant on improvement of production technology, plans will have to be drawn up and implemented to improve the transport system. This would require transport of coal by belt conveyors and locomotives, especially for mines having long wall faces. Other improvements that are required include better supports in mines, methods of dealing with fires and abnormal high heating, material supply systems, and rescue organizations. In the past four years, the government has been importing large quantities of sophisticated mining equipment. In the case of less-sophisticated equipment, capacity exists, and has been developed further, for manufacture within the country.

The organization and management of modern coal mining requires improvements in the areas of telecommunications and mining electronics. There are a number of aspects in mining in which electronics plays an extremely important part. This includes electronic mine safety devices and systems for monitoring and controling poisonous gases, air density, and moisture and for monitoring earth pressure and so forth. Recently, central dispatcher systems have been developed in India for facilitating the movement of materials within mines.

The importance of coal beneficiation in the context of India's vast resources of coal has been mentioned earlier. There is a trend all over the world to take advantage of lower transportation costs and fuel efficiency by cleaning run-off mine coal to the fullest extent possible. In most coal-producing countries, almost 50 percent of the total production is beneficiated, but in the case of India, only about 15 percent of the total production is washed. The problem of beneficiation in the Indian context is, therefore, really one of spreading existing

TECHNOLOGICAL ALTERNATIVES

technologies, while at the same time, improving the process of beneficiation in large-sized plants.

Some new technologies that could revolutionize the transportation and consumption of coal are receiving increasing attention. One of these is using coal slurry pipelines. Coal slurry pipelines carry finely ground coal in a suspension of water, with underground auxiliary pumping stations at intervals of 60 to 80 miles. Coal slurry pipelines can carry coal in this form to a distance of 1,000 miles (with capacities of up to 30 million metric tons of coal per year), based on existing technology. The quantity of water required for this purpose must be considered in evaluating this option, since this quantity is generally substantial. In those parts of eastern India where water supply is adequate, it may be possible to establish coal slurry pipelines. Oil is sometimes used as a substitute for water, and this leads to certain economic advantages, but the utilization of an oil-coal combination is a little more difficult than a water-coal combination.

There is a great deal of interest in India currently in coal gasification technologies. The Central Fuel Research Institute at Dhanbad and the Regional Research Laboratory at Hyderabad are both working on pilot projects to establish commercial application of coal gasification technology. There has been experimentation with a large number of coal gasification processes. Essentially, these consist of feeding in coal or lignite into a heated vessel with steam. The temperature in this vessel rises from about $1,000°F$ to as much as $2,000°F$ in a series of sequential sections. This causes the coal to be devolatized in the last temperature sections, and in this process, heat is produced. The carbon at this stage reacts with hydrogen at somewhat higher temperatures, according to the hydrogasification process. Then subsequently or concurrently, carbon and steam undergo a steam-carbon reaction. There are some aspects of coal gasification that are relatively unexplored and not fully understood. Empirical knowledge in gasification has been built up over the years, but a complete understanding of the rates of chemical reaction and the mechanisms that bring it about are not fully comprehended. Various processes are being experimented with all over the world. One of them is the high gas electrothermal process, in which coal and steam are injected in counter current directions. The Lugri process is relatively well known, and has been in existence for about 30 years. The overall gasification efficiency, which is a measure of the energy availability of fuel produced with this type of process, is about 95 percent. Approximately, 1 to 2 percent of the unburnt carbon escapes with the ash, and another 3 to 4 percent are lost in the form of heat from the total available energy. The overall Btu conversion efficiency for gas production is about 77 percent. There are other schemes for coal gasification that are well known. These include the Institute of Gas

Technology (IGT)-high gas-oxygen system, the IGT-high gas-iron system, the CO_2 accepter process, and so forth.

There are inherent advantages in the use of gasified coal, since it provides consumers with a clean and efficient fuel (unlike the direct use of coal itself). The use of coal in households nevertheless requires additional private investments in the form of pipelines within the household, metering, and appliances. The FPC estimated that this is likely to be of the order of Rs 650 per household and public investment in laying pipelines to the households in the region of Rs 500, wherever the density of the population is 10,000 people per square mile and with the assumption that every household uses coal gas. The costs of using coal gas would, therefore, be very high in comparison with some other sources of energy, like kerosene, bottled liquid petroleum gas (LPG), and so forth. It seems unlikely that in the near future, coal gas will be used on a large scale for domestic or industrial consumption, because in countries such as India, there is no demand for space heating, as in various colder developed countries. At the same time, it could be expected that if coal gas were to be supplied to a city like Bombay, which is located at a large distance from the nearest coalfield, there could be certain advantages in supplying coal gas by pipeline as against moving coal over the Western Ghat Mountains to Bombay. It may also be possible for a fairly substantial degree of substitution to take place from use of oil products to coal gas. Delays in implementation of research projects have been responsible for holding back the successful development of coal gasification technology in the country, but with the work currently in hand, it is likely that coal gasification will become commercially viable by the mid-1980s.

The possibility of production of synthetic crude oil (syncrude) and synthetic natural gas (SNG) is also subject to the same considerations as in the case of development of coal gas. The production of syncrude from coal requires efficient production of hydrogen and an upgrading of existing coal hydrogenation procedures. The South African Coal Oil and Gas Corporation has been producing approximately 4,000 barrels per day of high-grade gasoline and almost 100 barrels per day of good diesel oil, as well as some waxes from coal, at Sasol, South Africa. The process followed by the South Africans involves coal gasification, using a Lurgy reactor with oxygen and steam. This is then washed with methyl alcohol for removal of sulfur compounds and carbon dioxide. This newly formed gas is preheated and mixed with recycled gas, using powdered iron as a catalyst. Hydrocarbons are first condensed from the reacted mixture and then distilled to produce the desired petroleum products. Syncrude can be produced by a large variety of processes. Production was carried out on a significant scale during World War II, and a number of processes have been experimented with during the past three decades.

TECHNOLOGICAL ALTERNATIVES 111

A process that presents great potential and likely benefits in large measure is coal gasification in situ with air or other oxidizing agents introduced through a well drilled into the mine. (The history of this process goes back to the nineteenth century.) Successful methods of underground gasification of coal require either driving a shaft, using underground labor, or using shaftless methods, using bore holes to find access to coal deposits without using underground labor. The major control problems in this process are combustion control, controlling the roof, problems of lack of permeability, unexpected fracture, leakage, and contamination by water. Recent cost calculations indicate that for producing SNG by an underground process including transportation costs over a distance of 1,000 miles, it costs between U.S. $0.83 to U.S. $1.07 per million Btu. This makes the cost of this source of energy competitive. However, the process requires considerable development before it can be used on a large commercial scale.

In the chapter on energy resources, we have discussed the meager reserves of coking coal available in India. The problem of running out of these reserves in the future can be postponed by blending prime-coking coal with medium- and semicoking coal, as well as by adoption of new technologies in steel melting. While this would help in conserving India's coking coal reserves, it is essential to develop technologies for conversion of noncoking coal to other forms of coal, which could then be used in metallurgical processes. One process developed in the Central Fuel Research Institute, known as the "formed coke process," has shown promising results. Another benefit that could come from commercial development and application of this technology is that steel-melting capacity could be dispersed to different parts of the country, since India's reserves of iron ore and noncoking coal are relatively more widely spread than those of coking coal.

One could say in retrospect that the coal industry in this country has been subjected to prolonged neglect. The consolidation of the entire industry under one single organization may permit economies of scale in research and development work being exploited to their fullest extent. Since the coal industry is now in the public sector, research and development expenditures can be made as a matter of deliberate choice, based on social considerations rather than on a concern for immediate returns. The wide area that future research and development activities must cover pertaining to the coal industry requires large-scale investments and development of skills that the Ministry of Energy and Coal India can, perhaps, launch in the years ahead.

HYDROELECTRIC POWER

An assessment of the known potential for hydroelectric power generation, as presented in the previous chapter, indicates that considerable expansion is possible in the future in energy from this source. There are certain inherent advantages to electrical energy from hydroelectric sources.[1] First, hydroelectric power is not a resource that is likely to get exhausted since it is the result of a cycle based on solar energy, namely, evaporation, precipitation, and runoff. The environmental effects of hydroelectric power are generally less adverse than those of thermal power plants. The absence of a boiler, as in the case of thermal power plants, allows for a rapid start-up and loading; thus, hydroelectric plants are particularly suitable for meeting peaking demands. The operating and maintenance costs of hydroelectric plants are also found to be generally lower than those of thermal plants, with less frequent breakdowns than other types of generating equipment. In Western countries, hydroelectric plants are normally shut down for a maximum of two to three days a year due to unexpected breakdowns and about a week per year for scheduled maintenance. Thus, the total downtime of 3 percent (consisting of these two components) is almost one-quarter of the average downtime for thermal plants.

Disadvantages of hydroelectric power are its relatively high capital costs and inflexibility in location, since locations are specific to certain desirable geographical features (thereby frequently leading to higher transmission and distribution costs).

Various designs of turbines have been developed for application, depending on specific features related to the actual site of the hydroelectric power plant. Well-known designs are of the Pelton, Francis, or Kaplan types. The Pelton impulse-type turbines are generally installed at sites that have water heads of more than 1,000 feet. There is in existence a hydroelectric plant in Austria that uses Pelton turbines with an operating head of 5,800 feet. The Francis reaction-type turbine, on the other hand, is used at sites that have heads in the range of 100 to 1,000 feet. For low-water heads of up to 100 feet, fixed plate propeller designs of the Kaplan type are generally used. The highest head known to be used for a Kaplan is 290 feet at a location in Italy. In recent years, horizontal axial-flow bulb and tubular turbines have been developed for application at sites with heads between 15 to 50 feet. The development of these two types of turbines makes a number of sites favorable for development of hydroelectric power generation that were not possible with earlier conventional turbines.

The technology that would be most applicable for production of energy from hydroelectric sources depends on the size and location of the plant, the cost of the land and various compensations that have to be disbursed in order to procure it, and the social cost of relocation of highways, buildings, and other structures. Typically, the investment per installed kilowatt for hydroelectric plant is higher than for thermal power plants. The benefit of hydroelectric power lies largely

TECHNOLOGICAL ALTERNATIVES   113

in its low operating expenses, which are generally far below those of thermal electric plants. Estimates of capital costs for hydroelectric power plants in the United States were provided in the author's previous work.[2] Precise estimates of the investment per kilowatt for hydroelectric plants in India are not available, mainly because most hydroelectric projects are designed to provide a large range of benefits, which makes it difficult to estimate the hydroelectric component of costs and benefits. In most cases, they are generally found to be in the range of Rs 3,000 (and above) per kilowatt.

ELECTRIC POWER GENERATION

In terms of the total outlay on research and development, developments in nuclear power generation acquire priority. The technology of nuclear power also acquires importance in view of India's (known) scant resources in petroleum and hydrocarbons. A look at the resources of uranium indicates that about 8,000 megawatts of nuclear power capacity can be fed with nuclear fuel for a period of only 30 years. The entire strategy for development of nuclear power, as mentioned earlier, rests on the possibility of developing breeder reactors to enter the energy sector on a commercial scale in the future. The first-generation nuclear plants installed in the country consume uranium as fuel-producing electricity, as well as plutonium. Plutonium is required for feeding into the reactor of the future, namely, the FBR.

The economic advantages of nuclear power development were investigated by Kirit Parikh, using a multiregion, multiperiod linear programming model to evaluate alternative strategies for development of nuclear power.[3] The planning horizon used for specifying the objective function in this model was a period of 20 years, during which it was specified that power would be produced to meet demand at the lowest cost possible to the nation. His analysis indicated that nuclear power is economically justified and that the decision for setting up Candu-type reactors is economically wise. His analysis showed that the cost of delaying the development of the FBR may not be large, but it indicated that after 1990 or thereabouts, nuclear power produced in the country would have to come from fast breeders. The reliance on fast breeders would go on increasing in the period beyond this century. His results showed that if the FBR technology could not be developed, a total of 18,000 megawatts would have to be based on coal by the year 2000/2001. The magnitude of the problems involved in coal mining and transportation with this scenario can be visualized from the fact that a total of 60 million metric tons of coal would be required merely for feeding the coal-based power plants. The importance of developing the FBR can be visualized from Parikh's analysis, particularly in view of the vast reserves of thorium that are available in India. Even though progress in nuclear power research and development is not accessible to the public at large, an impression has been gaining ground among the concerned public in India, as well as among officials working in the nuclear field, that the original momentum in this area (generated in the early years of independence when Bhabha established India's nuclear program) has slowed down in recent years. Progress on the FBR, however, appears to be satisfactory, but unless adequate support and facilities are provided for its development on a commercial scale, this vitally important technology may not materialize in India before the next 15 years or so. Though the strategy for nuclear power development described in the previous chapter appears sound and logical, the tempo of developments must be upgraded in order to sustain the initial impetus.

Another area for power generation by sophisticated methods appears to be in the field of magneto hydro dynamics (MHD). MHD deals with the flow of an ionized gas flowing in a direction that is perpendicular to an existing magnetic field. When this happens, an electric field is induced in the direction that is perpendicular both to the gas flow, as well as the magnetic field. Electrons and ions in the gases permit a flow of electric current under the influence of the induced electric field. The effect utilized in MHD is often referred to as the "Faraday effect."

Ionized gases can be produced by direct heating, using combustible products, often facilitated by introducing easily ionized metals, particularly alkali metals, in the gas flow. The extent of ionization determines the electrical conductivity of the gas. The losses in conversion in the MHD cycle are produced because of frictional losses as a result of gas flows and because of the existence of nonhomogenous electric current flows arising out of eddy currents and so forth. Because higher temperatures occur in the MHD channel than in conventional steam generators, and combustion gases are directly used in the channel, theoretically, MHD can provide a thermal efficiency of about 50 percent on its own and about 60 percent when the MHD unit is combined with a conventional steam generation system. This compares very favorably with the conversion efficiencies obtained in fossil fuel plants, which are generally in the range of 30 to 45 percent. Major research efforts are being made to construct MHD converters, using oil, gas, or coal as the major fuel source. Problems being encountered in these efforts pertain mainly to the performance of electrodes and their life in service. Major problems in MHD programs are also being faced in developing suitable metals. For instance, the MHD generator requires a very high electron concentration in the gas, which can be achieved at very high temperatures of 4,000 to

TECHNOLOGICAL ALTERNATIVES

5,000°F. It is, therefore, necessary to develop materials that will function satisfactorily at such high temperatures and yet retain their corrosion resistance. Substantial progress has been made in development of MHD in the United States, the United Kingdom, Japan, and the Soviet Union. The Soviet Union has already taken a lead by setting up a plant of 25-megawatt capacity, using natural gas as the fuel. The National Committee on Science and Technology of the government of India has proposed a ten-year program for the development and application of MHD in India. This consists of the following phases:

1. Conduct experiments at the laboratory stage, using a 2-megawatt MHD generator for developing essential scientific and technical skills necessary in this process;
2. Design and set up a 25-megawatt plant, using MHD process; and
3. Develop design capability for a full-scale commercial plant, with a capacity of 500 to 1,000 megawatts.

Preliminary estimates indicate that the cost of developing MHD may be extremely high. Keeping these factors in mind, the FPC mentioned that efforts in this field should be subjected to constant review and evaluation of costs "before embarking on installation of higher capacity MHD units."[4]

Other processes are also available for increasing thermal efficiency of power generation, such as a combined gas-turbine system, using coal gas of low calorific value. BHEL already has a plan for development and application of this combined cycle power system, based on know-how already available in other countries. The combined cycle can be understood by the example of an existing combined cycle plant in operation in Lunen, West Germany.

Figure 5.1 shows the flow diagram of the process involved in power generation at this plant. Coal is first gasified in the gasifier, and ash as a by-product of the process is removed. The media used for gasifying are steam and air. Flue gas produced in the gasifier passes through a scrubber and then enters a pressure-dropping turbine. It is then fed to the combustion chamber of the boiler with a pressurized furnace. The combustion gas from the boiler then enters the main gas turbine, which produces about 74 megawatts of power potential. The exhaust gases from the gas turbines are used for feedwater preheating in the feedwater preheater. Coupled with the gas turbine is an air compressor. The air from this compressor is apportioned between pure combustion air for the boiler and gasifying air for the gasifier. Steam generated in the boiler drives the conventional steam turbine, and the gasifying steam is bled from an extraction stage of the steam turbine.

FIGURE 5.1

Combined Cycle Generation of Power

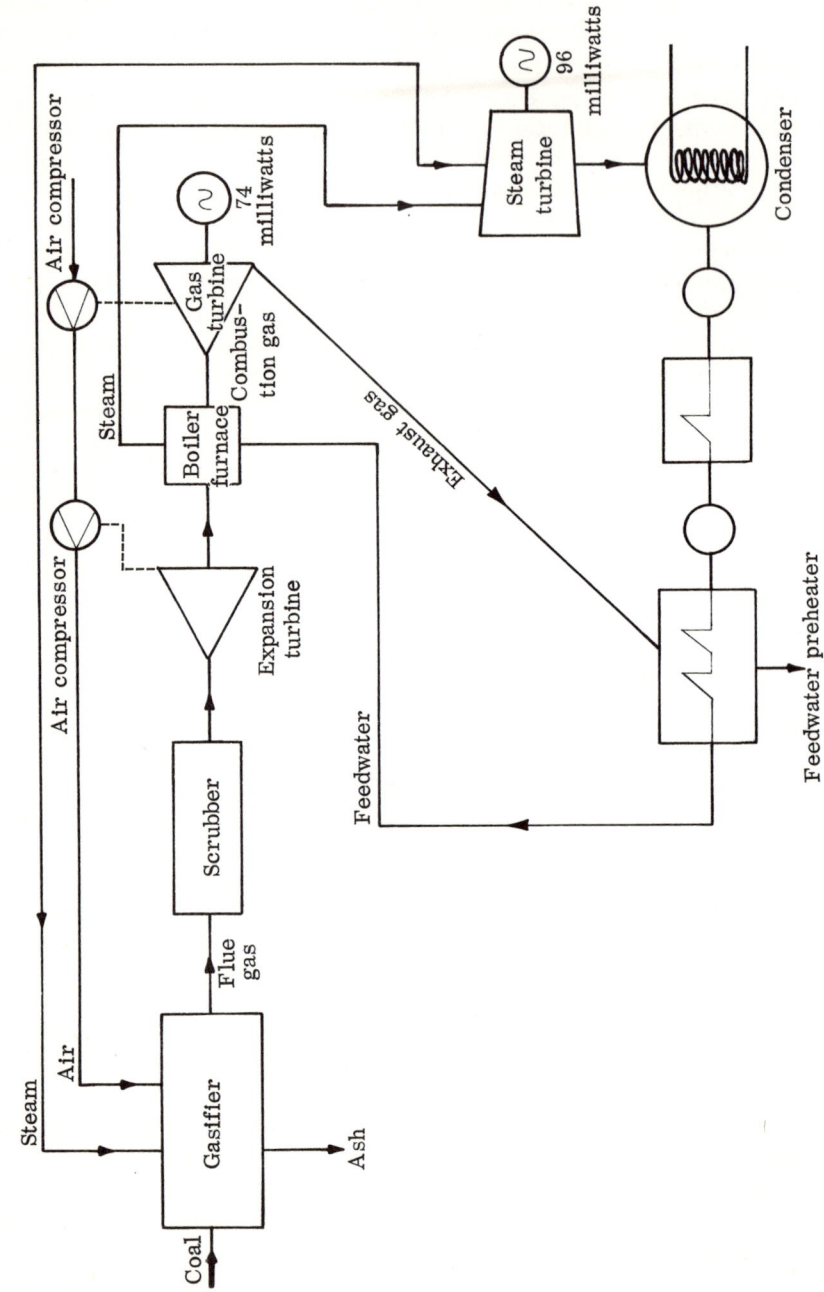

Source: Compiled by the author.

The Lunen plant has a total gross capacity of 170 megawatts, made up of 96 megawatts from the steam turbine and 74 megawatts from the gas turbine. The plant's efficiency is around 36.9 percent, corresponding to 2,330 kilocalories per kilowatt hour, as related to gross generation. Allowing for part supplies, the heat rate of the plant is approximately 2,390 kilocalories per kilowatt hour. It is expected that the combined cycle will be used for power generation in India sometime by the mid-1980s.

Another technology being developed by BHEL relates to the use of a fluidized bed for power generation. This process involves combustion of coal at temperatures between 800 to 900°C in a fluidized bed, operating at an elevated pressure and in direct contact with heat transfer pipes. The energy in the hot products of combustion is utilized for driving both gas and steam turbines. One benefit from this process is its ability to fix sulfur and prevent air pollution. The efficiency of fuel utilization with fluidized bed technology generally increases by about 5 percent, that is, the fuel requirement to produce an equivalent amount of power is reduced by 6 to 16 percent in comparison with a conventional thermal power plant. Fluidized bed technology also has special advantages for India, since it would permit utilization of the vast reserves of low-grade coal in the country.

NONCOMMERCIAL AND NONCONVENTIONAL SOURCES OF ENERGY

Solar Energy

In evolving strategies for the development of technological alternatives in solving the energy problems of the country, it must be borne in mind that almost half of the energy consumed in India comes from noncommercial sources. The problem of energy and development, is therefore, not merely one of substitution, by which least-cost solutions providing for a desired level of service are offered to society at large, but one of extending use of commercial energy into the rural sector, where options and opportunities are restricted due to the nonavailability of various forms of energy. The easy availability of solar energy in most parts of this country renders this particular form extremely relevant to development (and application over a wide variety of uses), particularly for rural India.

Solar energy can be used in a variety of ways. It can be used for heat raising, that is, for solar thermal systems; for conversion into electricity, that is, solar electric systems; or for producing fuel, either through photosynthesis, which can be achieved by large-scale solar plantations, or direct generation of hydrogen from water (known as "photolysis"). It is also possible to convert existing photosynthesis products, such as trees, plant waste, and animal and human waste, into fuel or to burn them directly. In a sense, the use of forest energy and conversion of waste into fuel, such as in biogas plants, is really an extension of the use of solar energy (as is every other form of energy on this plane for that matter). In this section, however, we will confine ourselves to discussing some of the more promising methods by which solar energy can be used directly rather than the process of photosynthesis or conversion of waste into fuel. The merits of solar energy development are based not merely on the magnitude of its availability but also on certain tentative but important economic considerations. For instance, the present cost for photo voltaic converters is more than a few hundred dollars per watt, which is higher by at least two orders of magnitude than competitive solar power utilization in central power stations, but lower costs for the former appear within reach.

One of the most attractive areas in which encouraging results have been obtained is the direct use of solar energy for heating purposes. For over 20 years now, the use of solar energy for production of hot water has been commercially proved and exploited in such countries as Japan and Israel. The increase of the price of petroleum has made solar-based hot water production and solar heating of space an economically viable proposition, even in the United States. The low levels of income and the existing climatic conditions in India make the use of solar-based hot water heaters less useful than in Western countries. But the possibility of using solar energy for heat-raising purposes in industrial hot water processes is an attractive and viable technology, and it would certainly contribute to a reduction in the demand for conventional fuels by industry. Simple examples are the use of solar energy for sterilization of bottles in a dairy and preheating water for industrial steam raising in a large range of industries, such as textiles and chemicals. The production of heat to raise the temperature of water up to 90 to 100°C can be achieved by the use of simple concentrators. A typical nonfocusing type of solar collector is shown in Figure 5.2.

Solar thermal systems can be used on a large scale for irrigation pumps. An external combustion engine where heat is provided from external sources to the cylinder with a working fluid inside is an ideal candidate for application of solar heating. A large number of such pumps have been constructed, including an engine that runs on hot air and an engine that runs on a low-vapor pressure fluid, such as methylene chloride (the vapor-pulse engine), which has no moving parts at all. The latter type of engine is particularly simple in design; a representative example of its broad features is shown in

FIGURE 5.2

A Typical Solar Collector

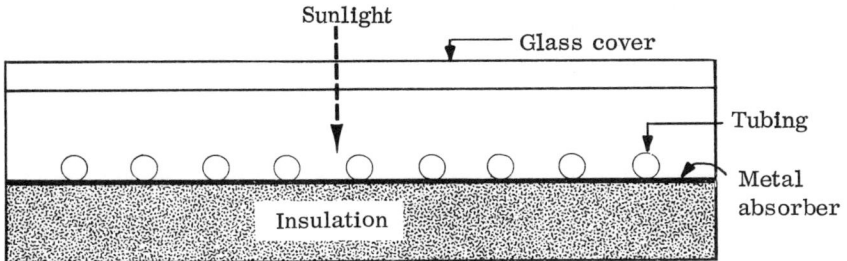

Source: Compiled by the author.

Figure 5.3. It is important to note that the efficiencies of a solar thermal engine that relies on a dispersed form of solar energy are naturally very low. Given the high cost of equipment that is required for utilization of solar energy in such cases, the costs of such application are still necessarily high. The key to improvement in the economics of solar energy used in solar thermal systems lies in better collection of the heat and in reduction in the cost of equipment required for utilizing it. The main reason why these costs are high is that without light concentrators, the maximum temperature that can be attained for the working fluid will generally be below 100°C. Hence, by the second law of thermodynamics, the efficiency of the entire operation is likely to be less than 5 percent or so. In order to produce large enough outputs of energy for application, both for domestic as well as industrial uses, the total size and scale of the equipment required would be naturally high, and, therefore, the cost of capital per kilowatt (using such systems) becomes high.

Solar thermal systems present a good potential for air conditioning as well. The principle used for air conditioning based on solar energy is the same as that of the traditional gas-heated refrigerator. In this case, solar energy provides direct heat to a refrigeration unit based on the vapor absorption cycle, where the fluid used can be ammonia or any other low boiling point fluid. In rural areas, there is considerable scope for making use of solar energy for refrigeration based on the vapor absorption cycle, particularly for refrigerating

FIGURE 5.3

Vapor-Pulse Engine

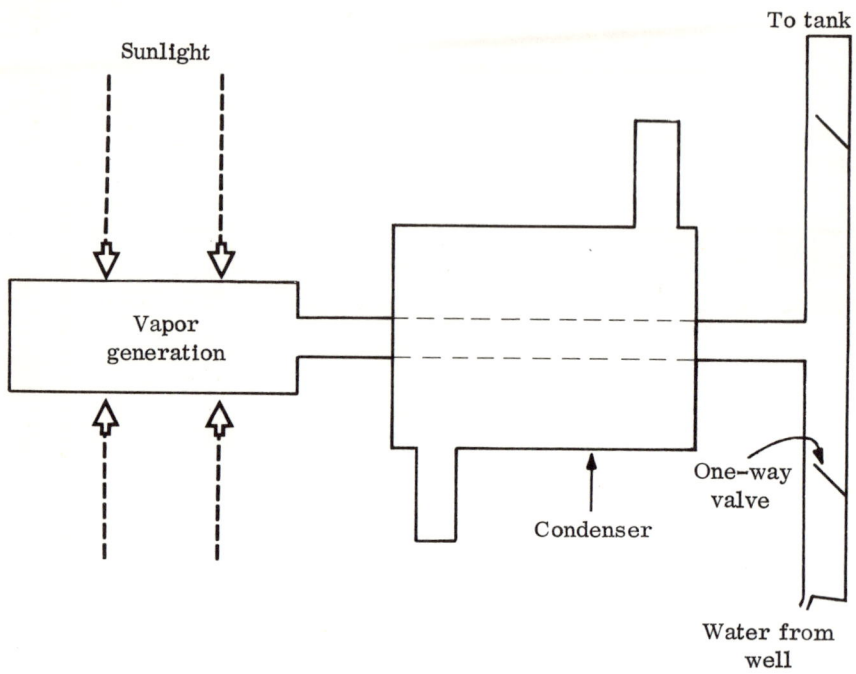

Source: Compiled by the author.

milk and other products that may normally deteriorate in storage without refrigeration. The weather conditions prevailing in India over most of the year also render this form of application attractive for air conditioning residential and office space. It is well known that even though the existing costs of air conditioning make this option a luxury, efficiency in various activities is generally promoted by the presence of an air-conditioned atmosphere. Air conditioning is also essential for a large number of industries where precision measurements have to be made and parts manufactured to close specifications and where tolerances can only be achieved under carefully controlled temperature and atmospheric conditions.

A simple schematic diagram for a solar space and water-heating system is shown in Figure 5.4. The system consists of a solar collector of reasonable size facing in the southward direction. This

FIGURE 5.4

Solar Space and Water-Heating System for a Typical House

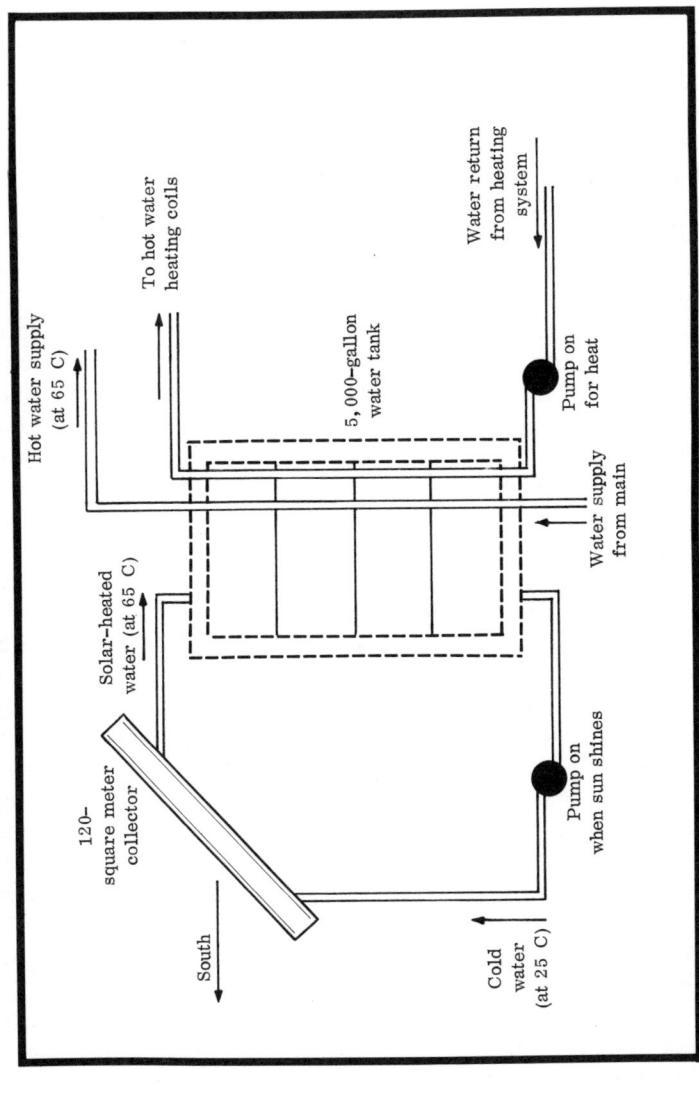

Source: S. S. Penner and L. Icerman, Energy, vol. 2: Non-Nuclear Technologies (Reading, Mass.: Addison-Wesley, 1975), p. 337.

collector heats entering cold water from 25°C to about 65°C in a suitably designed heat exchanger. The hot water that is produced is stored in a 5,000-gallon tank, which serves as a heat storage sump for the hot water space-heating system and the running hot water supply. The water supply is, of course, introduced from the water mains and is heated on passage through the storage tank before it is used. Water is then pumped through the heat exchanger solar collector during hours of sunshine. The primary water-heating, storage, and solar energy collection unit can be seen as being completely enclosed. Space heating, as opposed to water heating, is accomplished, using the same system, by means of hot water coils, which are embedded in the walls, floors, or other parts of the building structure.

Another application of solar energy that has large potential and on which work has already been started in India is the use of solar thermal systems for power generation. Solar energy can, of course, be converted directly into electricity, using semiconductor devices known as "solar cells." The use of solar cells is fairly widespread, and in particular in the space industry, solar cells have been used for a number of functions, particularly in space vehicles and satellites that have been sent up in recent years. A solar cell essentially consists of a semiconductor junction that absorbs the protons of light falling on it, and these are converted into electrons, which are collected, giving rise to electricity. This form of conversion of solar energy is known as "photovoltaic conversion."

Until quite recently, photoelectric cells were fairly expensive. Even now, a photoelectric cell may cost anywhere between Rs 50,000 to Rs 100,000 per kilowatt. This, when compared with the capital cost of, say, a hydroelectric plant, is exorbitant. Therefore, solar cells have not been regarded as a feasible alternative for conventional power generation. Interest has, however, been stirred in the last two or three years, particularly on account of the work done in the United States by two independent laboratories. Estimates indicate that the costs of solar cells based on mass production methods may come down to below Rs 5,000 per kilowatt. The concept of low-cost solar cells is quite simple, and the explanation of their tentative design lies in the fact that if we throw greater light on a solar cell, it will produce a larger quantity of electrons and, therefore, a larger amount of power.

What has been done in the two laboratories in the United States working in this field is to demonstrate that solar cells can be operated at 1,000 times solar intensity, thereby providing 1,000 times more power per square centimeter of solar cell. Researchers have used existing semiconductor technology, and in fact, one laboratory has even used the most basic semiconducting material, namely silicon. Both these laboratories have exhibited that by using simple lenses,

TECHNOLOGICAL ALTERNATIVES

proper heat storage sinks, and simple devices for attracting the sun as it moves from east to west, high-power outputs can be obtained throughout the day. Of course, by producing 1,000 times more power per unit area of the solar cell, the cost of the solar cell could perhaps be reduced to Rs 100 per kilowatt of power produced; the cost of the tracking system to focus on the sun, as well as the lenses, heat sink, hurricane proof foundations, and so forth, make up another Rs 2,500 to Rs 5,000 per kilowatt. The scheme outlined above can be used perhaps in the not too distant future on a fairly large scale in India.

This scheme is also of particular relevance because it does not require construction of large-sized power stations, as are being built nowadays to take advantage of economies of scale. Perhaps a large number of small generators could be built, which could be used for a variety of agricultural and other rural applications. For instance, the typical input required for an irrigation pump is five to ten kilowatts, and if the generation of power using such solar-based systems could be decentralized, then perhaps a large number of such solar stations could be located all over the country to provide power for irrigation purposes at competitive prices. It is also interesting to know that the time when irrigation is needed in large quantities—that is, from the months of October to March—over most of the country, the sun provides a large input of energy, which could be used for irrigation purposes, using these systems. It is also possible to combine such systems with charging of batteries for nighttime use, thereby providing illumination in villages all over the country based purely on solar energy. Undoubtedly, there would be a number of problems in the establishment and operation of solar-based thermal plants, particularly in remote rural areas. It would, however, be useful to set up some such plants in different parts of the country to gain familiarity with the problems associated with their establishment and operation.

Another process that can be used for generating power based on solar energy involves vaporization of a low boiling point liquid, which is then allowed to run a low-pressure steam turbine. Again, a limitation is imposed on the efficiency of the system by the low temperature that is achieved using direct solar thermal boilers. The cost of manufacturing large-sized turbines to take advantage of low-pressure cycles will continue to be high, and it appears that this method will not have a very wide application and economic benefit in the future. Therefore, we will not concentrate our efforts on developing low temperature systems.

Again, there are large benefits in concentrating sun rays on a small surface of collection from a larger surface. As far back as 2,500 years ago, Archimedes used this principle when he set the Persian fleet on fire in the harbor of Athens by focusing the rays of the sun on the fleet using the shields of the Greek army as mirrors. The same principle can be used for industrial purposes. Work in this area tentatively suggests that steam at temperatures of 400° to 500° C can be used on a commercial basis by focusing heat from the sun on properly designed boilers. This would naturally bring about much greater efficiency in solar energy conversion and power generation and also bring down the costs of turbines considerably.

The Sandia Laboratories in the United States are building a one-megawatt power plant based on this concept in the state of New Mexico. Designs for a ten-megawatt plant are also ready, and once funding is available, this project will be started.

The basic design of this system is quite simple. A large number of mirrors are located all around a tall steel tower, on which a boiler is located. This set of mirrors is placed in such a way as to form a shallow parabolic dish, which reflects the parallel rays of the sun onto the boiler, which is placed at the focal point of the parabola. These mirrors track the sun as it moves from east to west through a simple motorized arrangement. This enables the sun to always remain focused on the boiler over a period of eight to ten hours a day, and the boiler then generates steam, which is used for running a turbine. For a one-megawatt plant, achieving 25 percent efficiency, the total area of the mirror field is around 4,000 square meters.

There are many areas in the country where power can be produced for a number of days in the year. Western Rajasthan, northern Gujarat, and Kutch provide geographic situations where sunlight may be directly available for almost 330 days a year. BHEL is currently trying to set up a ten-kilowatt plant based on solar energy. Once this plant has been established, efforts will be made to set up larger plants and also borrow on the technology developed in other countries to extend the application of solar energy for power generation purposes in this country.

Experts and technologists who are skeptical about the application of solar energy on a wide scale generally point to the problems in storage of solar energy (solar energy is directly available, of course, only during daytime hours). Undoubtedly, this is a limitation that inhibits the growth of solar energy application, but it must be remembered that a large number of applications are such that perhaps energy need not be produced over the 24 hours of the day. For irrigation, for instance, storage of energy is not really necessary, since farmers irrigate their fields generally during the daytime, unless they are compelled by other reasons to do so during the hours of the night. There would, of course, be occasions when irrigation is required without the availability of sunlight. In such cases, it may be possible to get over the problem of nonavailability of sunlight by installing storage tanks of reasonable capacity at adequate heights to provide irrigation when required.

For large-scale power generation based on solar energy, the easiest way of storing energy is perhaps to use pumped storage plants. In pumped storage plants, water is pumped uphill at periods when surplus energy is available and permitted to flow downhill when the stored energy is to be used for useful purposes. India already has planned large pumped storage schemes, of which some are under construction in order to meet peaks in the load curve of the system. This will permit large generators to operate at almost full load during most periods, and hence, pump storage schemes have a natural advantage for application where large variations in load are imposed on the power supply system. For a pump storage plant, two reservoirs are generally required, one at an elevation, such as on a mount, and one at a lower elevation. The height of the former can be anywhere from 50 meters and above. On current indications, it appears that power generation using solar energy can be developed for applications where peak loads exist. The use of solar energy power generation in conjunction with pumped storage schemes is also a possibility that is likely to be developed in the future.

Another form in which solar energy can be used is through the use of ocean thermal energy conversion (OTEC), that is, by using the temperature difference between different levels of the ocean for power generation. Work is already in hand in the United States, and the Energy Research and Development Agency (ERDA) hopes that by the early 1980s, an experimental prototype OTEC plant would be available for generation of 25 megawatts. By 1985, ERDA's schedule indicates a commercial plant operating at 100-megawatt capacity.

The concept of OTEC goes back to the middle of the nineteenth century. A French physicist by the name of d'Arsonval predicted that electrical power would be extracted from temperature differences between the warm waters of the surface of the oceans and cool waters a few thousand feet below. The idea was later taken up by a French student of d'Arsonval, and he set up a small OTEC plant, which generated 22 kilowatts of power. Very soon after its erection, the plant was destroyed by heavy weather in the seas. Interest was seriously revived in OTEC technology in the 1960s.

The basic concept of the OTEC principle requires surface waters of about 27°C being pumped through tubes in an evaporator. These tubes are situated inside the working fluid, which has a low boiling point. This working fluid is vaporized due to the heat from the water, and the vapor expands through the turbine, which turns the generator. After performing this work, the expanded vapor leaves the turbine at a lower pressure and is channeled into a condenser, which is kept cold by deep ocean waters at temperatures as low as 5°C. The low temperature converts the vapor into liquid once again, and a pump returns the working fluid to the boiler evaporator to begin the cycle

FIGURE 5.5

The Ocean Thermal Energy Conversion Cycle for Energy Production

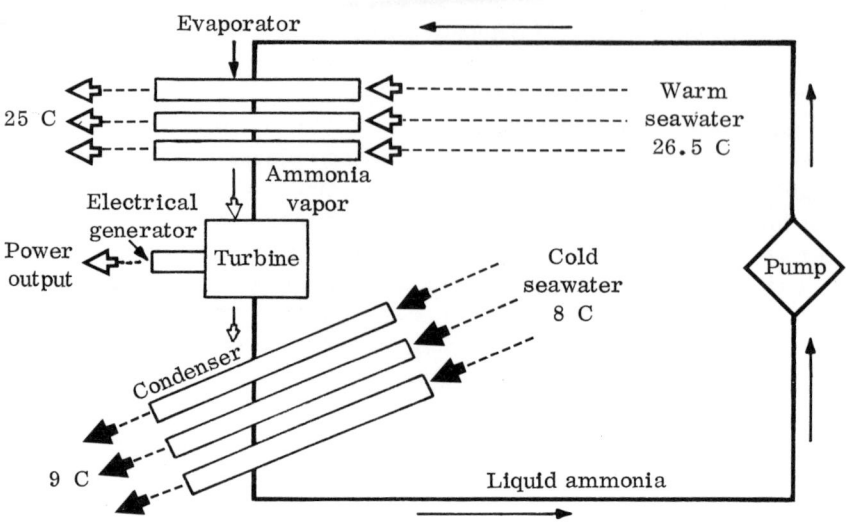

Source: Compiled by the author.

once again. Figure 5.5 shows a diagram of the OTEC cycle. With the large coastline that India possesses and the existence of fairly high temperatures on the surface layers of the oceans surrounding the country, application of the OTEC principle becomes relevant. Unfortunately, official agencies in India have not yet shown any significant interest in the application of this technology, and unless positive decisions are taken to develop OTEC-based power generation, India is likely to stay behind in this technology in the years to come.

In evaluating the potential for application of solar energy on a commercial basis, one must investigate some general features relating to the utilization of solar energy. It is essential that in the ultimate analysis, whatever form of solar energy application is developed, it must be competitive with other forms of energy. Solar energy converters would have to be constructed in such a way as to involve a capital cost that is in no case greater than the sum of the fuel and capital cost for fossil fuel-based energy systems. Since fossil fuel and plant capital costs are generally comparable, solar energy sources may be economically viable if their capital costs are double those of coal-, oil-, or gas-based systems, that is, in the region of U.S. $600 to U.S. $700 per kilowatt at existing prices. Using a rule of thumb, perhaps solar collection systems alone could cost up to U.S. $350 per kilowatt and the conventional turbine generator installation or other equipment for conversion of solar energy into other usable forms another U.S. $350 per kilowatt. In comparison with nuclear power generation, one could say that since nuclear fuel costs range between 10 to 15 percent of electricity produced in nuclear power plants, solar energy utilization systems can be introduced at capital costs that are 1.15 times those for nuclear plants. This means that solar energy systems would be viable at existing competitive prices if they were in the region of U.S. $1,000 per kilowatt (in capital investment). If we allow for inflation of nuclear fuel prices in the future, we may assume that solar collectors would be viable, that is, they would compete with nuclear power systems, if they cost something in the region of U.S. $500 to U.S. $600 per kilowatt.

In the immediate future, it seems unlikely that solar thermal collection systems will become commercially available at the prices indicated above. They are also not likely to achieve efficiencies of more than 20 percent in the immediate future. It appears, therefore, that a large-scale application of solar energy is not exactly around the corner and, at best, may become a viable alternative sometime toward the end of the century. There are, of course, hazards in "guesstimates" of this kind, since technological innovation and breakthroughs cannot be predicted with any measure of certainty. It is quite possible that the development of some kind of solar collection system will make the application of solar energy a reality much earlier than presently anticipated.

Biogas Energy

Another form of energy that has received considerable attention in India is the use of biogas plants.* This form of energy production has a fairly long history in India. It is generally known that the first anaerobic digester in the world was set up in India as early as 1900, but this development was largely ignored until 1937, when a sewage purification station was set up at Dadar in Bombay. The anaerobic sludge digester of this plant stabilized the highly offensive sludge that settled down from the raw sewage into odorless, stable humic substances. During the digestion process, a large amount of combustible gas (methane and carbon dioxide) was also produced. The gas was

---

*These are commonly referred to as gobar gas plants.

used for running a five-metric-ton Leyland garbage disposal truck. This truck used to run for 55 miles on four cylinders containing 1,400 cubic feet of washed gas at a pressure of 3,000 pounds per square inch. The stabilized sludge is rich in nitrogen and, therefore, makes useful organic manure; it was sold by the municipality to farmers in the region.

Since the Dadar plant was set up, developments have taken place in a number of locations and institutions. Actually, a great deal of research is being done on biogas plants, pioneered mainly by KVIC. The KVIC has a research center at Borivli near Bombay, where extensive research is being undertaken on various aspects of the problems connected with the design, manufacture, and operation of biogas plants.* Other agencies that have done useful work in this field are the Central Public Health Engineering Research Institute, now known as the National Environmental Engineering Research Institute, and the Uttar Pradesh government Planning Research and Action Institute, which is located close to Lucknow.

Proponents of biogas plant technology draw a parallel between nature's oldest method of disposing of waste, that is, by decay or decomposition through the medium of various tiny microorganisms (bacteria). Anaerobic digestion in controlled biogas plants is merely an imitation of the natural process of anaerobic digestion, which takes place whenever organic matter decays under natural conditions.

There are two types of digesters in use today. The first is the batchfed digester, which is filled all at once, sealed, and emptied when the raw material has stopped producing gas, and the second more frequent type, known as the "continuous fluid digester," which is fed a certain definite quantity of the biogas raw material at regular intervals, so that gas and fertilizer can be produced on a continuous basis. Figure 5.6 shows the layering of products in the digester.

Anaerobic digestion consists of three phases. The first is hydrolization, the second, acid formation, and the third, methane formation. Bacteria have cell membranes through which food has to pass directly, whereas soluble substances, which are not of large molecular weight, can pass through cell membranes through diffusion. Sludge particles and large polymers cannot, therefore, pass through such membranes. The process of hydrolysis is somewhat similar to that of the hydrolytic process that occurs when food is digested in the stomach of a human being. In the anaerobic digestion process, acid formation takes place when organic substances are oxidized to acids in a manner that is similar to the formation of formic, acetic, and other acids observed in decay of organic matter. Certain specific bac-

---

*Biogas is cow dung gas.

TECHNOLOGICAL ALTERNATIVES

teria convert the volatile acids produced in this manner into methane and carbon dioxide. A mixture of these gases is known as digester gas, or cow dung gas. This gas generally consists of 30 to 40 percent carbon dioxide and 60 to 70 percent methane. The continuity of the process is ensured because the bacteria are capable of rapid reproduction and are not highly sensitive to changes in the environment. Methane-producing bacteria, however, work best in temperatures between 95°F to 100°F. There is a fall in gas production at temperatures below 68°F, and in fact, the reaction stops completely at temperatures of 50°F and below. This sometimes imposes a serious limitation on the usefulness of biogas plants. In the northern part of India, where cattle populations are generally found to be quite high, the existence

FIGURE 5.6

Layering of By-Products in the Digester
(layers not proportionate to actual model)

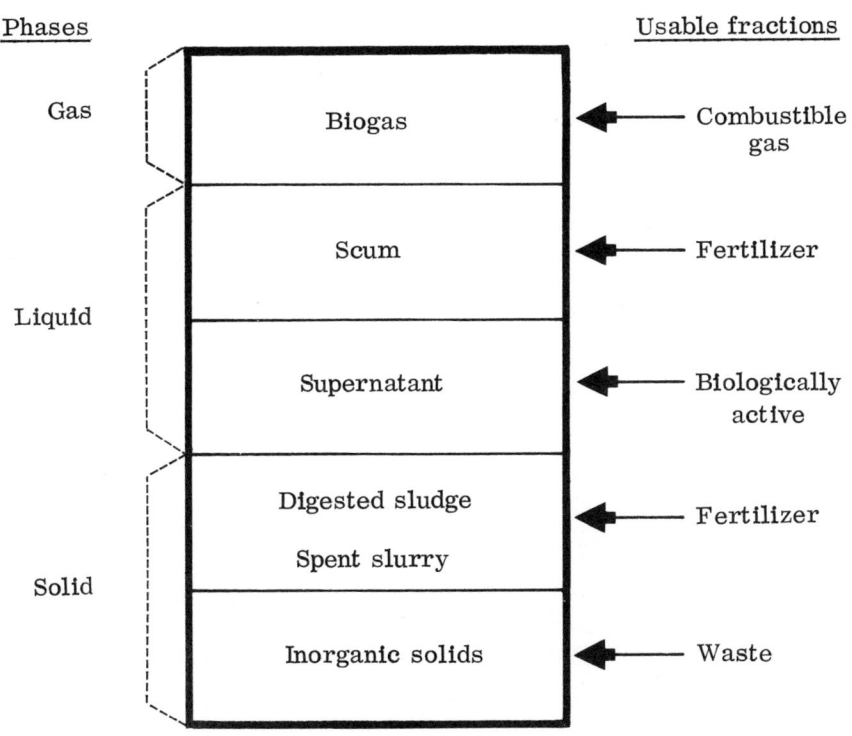

Source: Compiled by the author.

of low temperatures during the winter season reduces the possibility of using biogas plants for energy production. This generally calls for some external heating devise to maintain the temperature at 70°F and above for a reasonable rate of biogas production. The relationship between gas production and temperature is shown in Figures 5.7 and 5.8.

Both vertical and horizontal designs for biogas plants are available. The vertical design typically used in Indian conditions consists of a digestion tank with a minimum volume equal to the product of the daily input of dung-water slurry in kilograms and the number of days required for fermentation. When a tank of this type is covered and a vent pipe provided for exit of the gas, problems can arise. For one, a separate cost storage system has to be set up. Also, changes in the volume of fermenting slurry due to fluctuations in feeding the input or of removing the sludge for use as fertilizer can result in changes in the gas space, which may lead to entry of air, thereby reducing the heating potential of the gas. Maintenance and scum removal in this type of digester is difficult. To overcome the problem of fixed covers, designs have been established for using floating covers that can serve as gas holders: they ride up and down in accordance with changes in the level of the slurry at the base of the plant.

FIGURE 5.7

Gas Production and Temperature

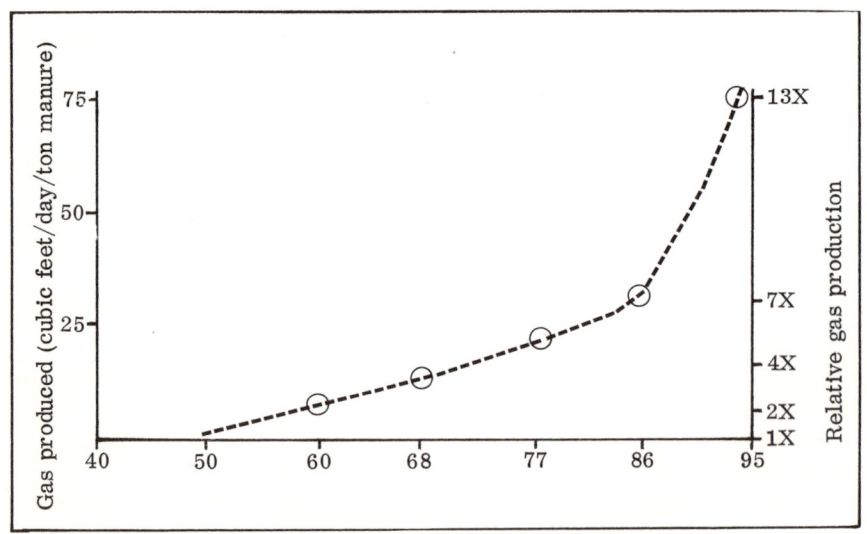

Source: Compiled by the author.

FIGURE 5.8

Comparison of Gas Production Rates at 60° and 95°F

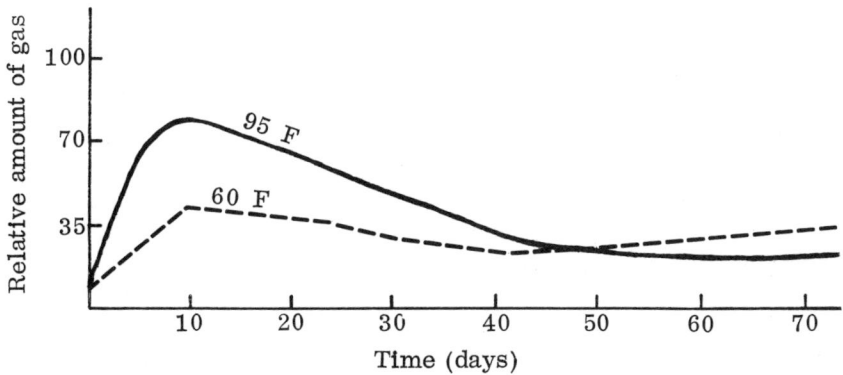

Source: Compiled by the author.

A typical horizontal design of a biogas plant consists of a long cylindrical tank placed lengthwise on the ground. The material that is used as input is loaded at one end and pushes against the previously loaded material, so that after a certain number of days, a digested sludge comes out at the other end. Certain advantages have been put forward in support of this type of digester. It is felt that the flow direction can be easily reversed if it is required for purposes of controlling the process. Also, the formation of scum takes much longer, and maintenance and cleaning are much easier with this design. This digester, nevertheless, still requires a gas storage system, and the possibility of air entry exists unless a flexible material is used for construction. Some recent installations have used long butyl rubber sausage-shaped digesters, which collapse on their own unless there is sufficient slurry and biogas inside to hold it up, much as a balloon requires air to hold it up.

A great deal of work has been done by researchers in the problems of designing, manufacturing, and operating of biogas plants. Amulya K. N. Reddy and K. Krishna Prasad, in various forums and publications, and in particular in a paper presented at the National Seminar on Energy in 1976, have put forward the view that small-scale family-sized plants do not go very far in solving the problems of rural energy in India.[5] Small-scale family-sized plants generally have a capacity of about 60 to 100 cubic feet of gas per day and require at least three heads of cattle for them to be running on an economically viable basis. These plants are estimated to cost between Rs 2,000 to Rs 3,000 per unit. Reddy and Prasad feel that these plants cater to the demand from cattle-owning high-income populations in the villages who do not account for more than 10 to 15 percent of the population in rural India. Using this argument they contend that it is unlikely that such plants will make a major impact on the energy problem of rural areas.

This author feels that small-scale family-sized plants certainly can help in solving energy problems in villages. First, even if this helps only 10 to 15 percent of the rural population of the country, to that extent, a fairly substantial reduction in noncommercial energy use can be brought about. Second, it is expected that the successful operation of such biogas plants would have an important demonstrator effect. It may be possible that small-scale farmers would then get together and share the costs and benefits of such plants on a cooperative basis, particularly since the operation, maintenance, and so forth of these small plants is certainly more manageable than that of large plants. It must also be emphasized that administrative, organizational, and other problems associated with the running of large-scale biogas plants are often quite beyond the capabilities of rural organizations as they exist today. There is also no doubt about the fact that private ownership of biogas plants would permit efficient and better utilization of these assets than is likely under large-scale and diffused ownership should larger plants be installed in rural areas. There are major problems in setting up effective collection systems, compensation systems, operating and maintenance systems, and distribution systems for biogas produced on a large scale under collective or large cooperative ownership. Even though the idea of economies of scale is very attractive, it is unlikely that this benefit can be exploited in reasonable measure unless major changes in the social and organizational structure at the village level take place in the near future.

Before biogas plants can become commercially attractive and be installed on a large scale all over the country, certain technical problems require urgent attention. First, the existing designs of biogas plants used in the country are not satisfactory. It is possible to evolve cheaper and more efficient designs for which cheaply available materials can be used. Once such designs are perfected, then mass production can perhaps bring down their costs of manufacture; it may then be possible to buy biogas plants off the shelf, with minimum fabrication and installation costs. It is also important, at the same time, to produce cheaper designs and materials for gas holders and to improve the efficiency of the burners, which are the primary appliances in which biogas is used in domestic consumption.

There are also various organizational problems that require attention. For instance, simple maintenance procedures can be es-

TECHNOLOGICAL ALTERNATIVES

tablished and information on these supplied directly to the owners of biogas plants in order to reduce the cost and dislocation whenever maintenance problems arise. There is also a need to improve the efficiency of the chemical process of biogas production. Research in this field can perhaps lead to the use of catalysts or other substances that may accelerate the process, particularly in lower temperature-operating conditions.

There is a wide range of applications in which biogas can be used, particularly in rural areas. These include the use of biogas for cooking purposes, for running engines for irrigation, and other applications. It can even be used for transportation. Biogas can be used for running tractors, power tillers, and other engines where petroleum products have been used in the past. Suitable engines can be designed for production of electricity directly from biogas as well.

Arjun Makhijani has worked out details of a complete village biogas electricity system in which biogas is used as the energy source.[6] His estimates for installation of such a system in the village of Mangaon, in India, which he studied, work out to U.S. $460 (with cooking fuel) and U.S. $265 (without cooking fuel). This compares favorably with the capital costs of other forms of generation of conventional type. Whereas studies of this type provide interesting indicators to what is possible, there are various difficulties in their implementation on a large scale. In particular, most people in India will harbor a social prejudice against the use of biogas, since it comes from waste matter. This certainly is a serious limitation in the use of biogas for cooking purposes. There are also problems associated with the operation and maintenance of any equipment in Indian villages, even if it is not of a sophisticated variety. As an example, even though biogas-based power generation systems may be economically viable, it would be difficult to support them in rural areas, where facilities for repairs, maintenance, and so forth are extremely limited and, in some cases, totally nonexistent. Under the circumstances, it is felt that other than for domestic fuel, in rural areas, biogas energy can at best be used for very simple applications for which technology and know-how is already in existence. Such applications would be in the area of lighting and heating in small-scale industries. Industries of the type where biogas could be used in rural areas include production of sugar and jaggery, brick making, pottery making, blacksmithing, and so forth.

Wind Energy

Even though India has strong winds in a number of parts of the country, the potential for application of wind energy in useful purposes is limited. This is due to one basic feature of wind energy that is found in most wind patterns all over the world, namely, the fact that the daily wind pattern in most places is extremely variable. Wind speeds and directions generally change over a wide range during different times of a given day, as well as from day to day. Periodic wind patterns are seen to recur from one day to the other, but the intensities of wind may vary considerably. As an example, certain coastline regions experience higher wind velocities during the day than at night. But the intensities could vary by large margins from one day to another.

On the other hand, the monthly average speed and direction of wind in a given area is surprisingly predictable. In contrast to the daily wind patterns, monthly average wind patterns vary by very slight amounts throughout the year, as well as from year to year. This high degree of predictability on a monthly basis can be explained partly by the fact that wind variations can generally be accounted for by conditions of meteorology, which have relatively short durations. Daily fluctuations, therefore, generally tend to cancel each other when monthly averages are computed, because variations generally take place on both sides of the mean. In most locations, monthly average wind velocities are rarely less than half of the annual average. Most monthly averages fall within a range of 15 percent of the annual average. The average annual wind velocity for a given location is understandably more stable than that of given monthly averages. For instance, typically, the annual average wind velocity in a given location will seldom fluctuate by more than 10 to 15 percent from year to year. The stability of monthly average wind patterns and long-term annual wind patterns is particularly relevant in utilization of wind power. The reliability of output of energy from a wind power system is naturally dependent on the stability of the average wind patterns, and the economics of wind power utilization are directly related to the consistency observed in average wind patterns. The capital cost of a wind power system is influenced by the cost of energy storage systems that would have to be established in order to take care of variations in wind velocity. The unit cost of energy generated from a given wind power system depends on the consistency of average wind patterns.

Various thousands of windmills have been used throughout history, but there appears considerable scope for improving the design for the future and in bringing about higher efficiency of utilization. The efficiency of a windmill is defined as the ratio of the power extracted from the wind to the power contained in the wind that passes through the swept air of the windmill itself.

There are various reasons why large-scale adoption of wind power appears promising and attractive. First, wind power is an ever-present resource of which there is no danger of depletion. Further, there is a fairly well-established base of knowledge, as well as well-known technologies that have been used in the past for use of wind power. The major disadvantage of wind power is the fact that the efficiency of utilization is extremely low when compared with other energy conversion processes. At the same time, capital costs are generally high, and there is a certain cost involved in terms of noise and pollution of the landscape that would exist if a large number of wind power systems were to be established on a widespread basis.

Wind power, unfortunately, is not a reliable energy source. The output of a wind power system depends on the wind velocity, and since wind velocity variations do not generally coincide with power demand cycles, it becomes necessary to provide for some form of storage device to take care of the mismatching between production and demand patterns. Three types of storage systems are feasible with wind power systems. These are pumped storage, compressed air storage, and synthetic fuel storage systems. It is widely felt that pumped storage and compressed air storage systems are not feasible for large-scale wind power systems; hence, the production of hydrogen by electrolysis and suitable collection of hydrogen for storage and ultimate use is the more feasible of the three alternatives. A storage system based on hydrogen gas storage would be particularly useful for wind power systems, because the hydrogen can either be used directly or transported from the site of the wind system to any other community, where it can be converted into electricity. The transportation cost of the energy produced by wind power systems is an important factor, because the high average wind velocities in the range of 20 to 30 miles per hour required for proper and efficient wind power utilization are not generally found in heavily populated areas. This, therefore, requires wind power to be generated and collected in locations far away from heavily populated areas; the hydrogen would then be transported to urban and other populated areas.

It is difficult to arrive at uniform estimates for wind power generation and utilization, mainly because there are a host of variables that must be assessed that are specific to a particular location. With a given wind power system design and a known investment cost, the cost per unit of the output from such a system is a function of the average annual wind speed at the site and the fluctuation of the actual wind speed from the annual mean. Investigators at the United Nations indicated in a study in 1957 that the effects on costs resulting from differences in the variables considered are substantially high.[7] The total wind power system cost, the power capacity system, and the cost per unit of the capacity as functions of the designed wind speed are shown in Table 5.1.

When looking at the figures given in this table it can be seen that designing a wind power system for the appropriate wind speed is of

TABLE 5.1

Relative Wind Power System Costs, Power Capacities, and Costs per Unit of Capacity as Functions of the Design Wind Speed

Design Wind Speed (mph)	Relative Total System Cost	Relative System-Power Capacity	Relative Cost per Unit of Capacity
35	1.00	1.00	1.00
30	0.86	0.63	1.36
25	0.71	0.36	1.97
20	0.57	0.19	3.00
15	0.43	0.08	5.37

Source: S. S. Penner and L. Icerman, Energy, vol. 2: Non-Nuclear Technologies (Reading, Mass.: Addison-Wesley, 1975), Table 15 6-1, p. 429.

extreme importance. The cost per unit of output for a wind power system design for 15 mile per hour winds is 5.37 times the corresponding cost of the system designed for 35 mile per hour winds. This cost, however, is not the only important basis for assessing the economics of energy generation in a wind power system, because reduction in the design wind speed affects the total achievable annual energy output per unit of installed capacity, which is known as the "specific output" and is measured in kilowatt hours generated per kilowatt installed.

The specific output is a useful indicator for a wind power system. Table 5.2 shows the relative specific output as a function of the design wind speeds for sites with varying annual mean wind speeds. It can be seen that the specific output increases with wind speed for all design wind speeds, and it is much larger for smaller design wind speeds. Looking at both Tables 5.1 and 5.2, it is also observed that a reduction in the wind speed from 35 to 15 miles per hour increases the cost per kilowatt more than fivefold, and the specific output is at the same time increased more than 24 times if the annual mean wind speed at the site is actually only ten miles per hour. These two tables also give us a basis for calculating the relative costs of energy produced from wind power systems with different design wind speeds specific to locations that have different annual mean wind speeds. The relative costs per unit of electrical energy generated in a wind power system are functions of the design wind speed and the annual capital costs associated with the sites where different annual mean wind speeds are

TABLE 5.2

Relative Specific Output of Wind Power Systems as a Function of the Design Wind Speed for Sites with Different Annual Mean Wind Speeds

Design Wind Speed (mph)	Relative Specific Outputs for Given Annual Mean Wind Speeds (annual kilowatt hours generated per installed kilowatt of electric power)		
	10 (mph)	15 (mph)	20 (mph)
35	1.0	8.0	16.1
30	2.9	12.4	22.2
25	6.3	19.1	30.1
20	12.4	27.2	39.3
15	24.2	37.8	48.1

Source: S. S. Penner and L. Icerman, Energy, vol. 2: Non-Nuclear Technologies (Reading, Mass.: Addison-Wesley, 1975), Table 15 6-2, p. 430.

observed. Values based on the above two tables, showing the relative cost per unit of electrical energy output for different wind power systems as functions of the design wind speed and the annual capital charges for sites with different annual mean wind speeds, are given in Table 5.3. It may be seen from these figures that when a site is subjected to mean wind speed of 15 miles per hour, the optimum value for the design speed is about 25 miles per hour. If the mean wind speed at the site is around 20 miles per hour, the design wind speed that is most economical for energy production is around 30 to 35 miles per hour. The variation of the unit energy cost with the design wind speed, which is near the minimum cost design, is small enough so that the precise value accepted for design wind speed is not of a great importance. It must also be mentioned that if a single design wind speed has to be chosen for a system that is expected to operate under highly variable wind conditions, then the most suitable values of the design wind speeds would appear to be close to 25 miles per hour.

The reason that wind power systems in India have not been used in the past is because these were either not considered feasible in most parts of the country or were not economically viable. It appears that the potential for development of wind energy applications in the country is limited. The FPC has, however, recommended that development of five- to ten-kilowatt wind electricity generators should be taken up in order to gain experience and knowledge in this field of application. It must also be emphasized that research and development

TABLE 5.3

Relative Cost per Unit of Electrical Energy Output for Different Wind Power Systems

Design Wind Speed (mph)	Relative Annual Capital Charges[a]	Relative Costs per Unit of Electrical Energy Output for Given Annual Mean Wind Speeds[b]		
		10 (mph)	15 (mph)	20 (mph)
35	1.00	1.000	0.125	0.062
30	1.36	0.469	0.110	0.061
25	1.97	0.313	0.103	0.065
20	3.00	0.242	0.110	0.076
15	5.37	0.222	0.142	0.112

[a] The same percentage of annual capital charges is assumed in all cases.

[b] These figures are obtained by dividing the relative annual capital charges by the value of the relative specific output for each wind speed.

Source: S. S. Penner and L. Icerman, Energy, vol. 2: Non-Nuclear Technologies (Reading, Mass.: Addison-Wesley, 1975), Table 15 6-3, p. 431.

for wind power systems should be directed toward evolving designs based on simple material and technology, which would not present major maintenance and operation difficulties once the wind power system had been erected at site. It is relevant to mention here that in other countries, notably in the Philippines, windmills have been built using bamboo, and these have been utilized for irrigation purposes.

Forest Energy

We have reserved discussion of energy from forests right until the end, since the problems associated with development of forest energy are not primarily technical in nature, but rather administrative and social. The lack of an effective forest policy in the country, and in particular the problems in its implementation, give urgency to evaluating our forest resources, as well as to evolving technologies and methods by which they can be used efficiently in the future. The easiest and most widely used method of obtaining forest energy is from the combustion of wood obtained from trees. Fuel wood is obtained from any tree.

The most important factor that can be controlled and influences the efficiency of wood as a fuel is the moisture content. (The gross calorific value of oven dry wood is around 4.7.) Moisture in freshly felled trees can amount to more than 100 percent of the dry substance, and therefore, this reduces its value as fuel, because of absorption of heat when water evaporates. It is not unusual to go to Indian villages and see specially cut moist wood being used for fuel, particularly for cooking in a village chula (hearth). This is an extremely inefficient method of utilizing valuable resources. The effect of moisture on the calorific value of wood can be seen from Table 5.4. Reducing the moisture content in wood is desirable for two basic reasons. First, it reduces handling and transportation costs, and second, it increases the fuel value of the wood. For efficient utilization of wood as a primary fuel, it is essential to cut trees three to four months before use in tropical climates and six to seven months in temperate zones. Of course, lack of proper storage facilities in a number of parts of India reduces the efficacy of this period of storage, since rainfall may add to the moisture content if wood is stored in the open.

Use of wood as a primary fuel is no longer favored to the same extent that it was in the past. There are various ways by which secondary forest fuels can be produced as a result of conversion of wood to more valuable fuels—by the process of carbonization, distillation, and air gasification. The most widely known method of upgrading the value of wood as a fuel is to convert it into charcoal. Charcoal is produced as a result of the chemical reduction of organic matter under carefully controlled conditions. It is widely used as a secondary source of energy and as a chemical agent in industry, but it also has a variety of other applications. This form of energy is particularly relevant to countries that have large amounts of manpower. As a comparison of the improvement in the amount of heat contained in a unit of charcoal, it may be mentioned that approximately 4.7 million kilocalories are contained in a metric ton of urban dry wood but 7.1 million kilocalories are contained in a metric ton of charcoal. There is a loss of about 2.6 million kilocalories (that is, 55 percent) per metric ton of wood, which is converted into charcoal, assuming a 30 percent yield of charcoal. In actual terms, however, this loss is below 30 percent, because air-dried fuel wood contains not more than 3.5 million kilocalories per metric ton due to the moisture it contains. No special energy is required for carbonization in kilns, since all the necessary energy is provided from the combustion of the wood that is charged. Therefore, it may be concluded that even though conversion leads to a decrease in the total available energy, its advantage lies in the more desirable properties of the products obtained.

TABLE 5.4

Comparison between the Percentage of Moisture Content and Calorific Value of Wood, Determined on a Dry Weight and Wet Weight Basis

Percentage Moisture		Calorific Value (kilocalories per gram)	
Dry	Wet	Dry	Wet
0.0	0.0	5.0	5.0
10.0	9.1	4.5	4.5
25.0	20.0	4.0	4.2
50.0	33.3	3.3	4.0
100.0	50.0	2.5	3.3
200.0	66.7	1.7	3.0

Source: D. E. Earl, Forest Energy and Economic Development (London: Oxford University Press, 1975), Table 3.4, p. 24.

Charcoal is generally made from wood, but it is also possible to use other materials, such as coconut shell and bone, which provide important and valuable charcoal, particularly for specialized application. In coastal South India, coconuts are produced on a large scale, and the conversion of coconut shell into charcoal is economically a viable alternative to using it directly for fuel. Dry wood is naturally a better input for charcoal production than wet wood, and the time required for carbonization is much shorter. Moisture can be removed by air seasoning wood before carbonization, and therefore, it is desirable to cut and split the material to the desired size when green to provide the largest surface area possible for evaporation. This cuts down the time for air seasoning. It is not essential, however, to use dry wood, for reasonably good results have been known to have been achieved with wood containing 100 percent moisture. Also the existence of moisture does not appear to affect the quality of charcoal produced, and sometimes, it may be economically desirable to sacrifice some charcoal yield for the advantage of a quicker return on the money spent on felling and storage of forest wood. It is also relevant to observe that dry wood is much lighter than wet wood; this makes it desirable to use the former, since a greater amount of labor is expended in handling wet wood. There are various techniques used in carbonization, but the basic principle common to all of them is that of heating the wood under carefully controlled conditions. There are, however, differences in the appliances used to achieve carbonization, which include kilns, retorts, continuous kilns, and furnaces.

Kilns have been known to exist from time immemorial. In kilning, part of the wood that is charged provides the energy needs to initiate carbonization. The principle involved in all kilns is similar and depends upon the combustion of part of a pile of wood until it is hot enough to be able to react exothermically in a limited air supply, that is, to bring about carbonization. In India, earth or direct kilns have been known to exist for a long time. The advantage of these is that handling of wood is reduced to a minimum, and little or no capital expenditure is required for the equipment used. Retorts are containers in which wood is subjected to heat; the heat is applied to the external surface of the equipment until the charge is converted to charcoal. Gases that are produced can be collected and, if condensed, may be fractionated into products, such as methyl alcohol, acetic acid, and pitch. This process is termed <u>distillation</u> when by-products are collected for commercial use. The advantage of using a retort as against a kiln depends on the market for by-products, so that the higher capital costs are offset by earnings that can be produced by selling the by-products.

Continuous kilns have been developed in a number of countries for continuous production of charcoal and are often referred to by mistake as "retorts." This equipment consists of a vertical steel cylinder, in which raw material is fed in from the top and charcoal is withdrawn from the bottom. In a continuous kiln, energy from an external source is required for at least part of the carbonization cycle. There is no recovery of by-products. The use of continuous kilns is restricted to specialized situations, where labor is available by day and night in order that a constant supply of wood can be fed into the kiln (as required in the process).

Distillation of wood has been used for centuries (the Egyptians, who recovered tar and acid from wood, used them for embalming purposes). In distillation, the wood charge is heated in a closed container arranged in such a way that gases and liquids that are produced pass through a condenser. Those gases that are not condensed can be utilized as energy sources on their own, and the net gases, as well as the tar, which is soluble in water, are collected from where they can be decanted and fractionally distilled to provide a range of useful organic chemical products. Some of the products and their applications produced as a result of distillation are gas for burning as fuels; methyl alcohol as an industrial solvent; acetic acid for conversion to acetone; wood oil, which is used for a number of chemical applications; pitch for road making; and so forth. In Australia, methyl alcohol produced from distillation is used as an aviation spirit (fuel), and acetic acid is used on a large scale in the food industry. Methyl alcohol is a very high octane fuel with antiknock properties, and this could be easily blended with gasoline in order to achieve substantial economies for transportation where consumption of fuel is considerable.

Gasification of wood has also been known for a long time. Two major products of gasification from wood are producer gas, which consists of carbon monoxide and nitrogen, and water gas, which is a mixture of carbon monoxide and hydrogen with small quantities of carbon dioxide and nitrogen. Producer gas is made by supplying insufficient air to burning carbon, which is converted into carbon monoxide. The reaction is exothermic, and therefore, quite often, the gas produced is used in furnaces immediately after production so that its heat absorption is not allowed to be wasted. If, however, this producer gas has to be used for internal combustion engines, it has to be washed, which takes away a large part of the heat. Water gas, on the other hand, is made by passing steam or a fine spray of water over burning charcoal. Water gas so produced is either burnt or used as a source of hydrogen, as in the Bosch process, as well as for synthesizing methyl alcohol. Producer gas and water gas are generally manufactured together so that the exothermic heat of the former alternates with the endothermic reaction in the second case. There is a large producer gas plant in operation in India in the Travancore region. The ash from this producer gas plant is sold to farmers of the region as fertilizer.

The development of superior alternatives in use of forest energy is imperative in the case of India. Since 50 percent of the energy consumed in India comes from noncommercial sources (this mainly applies to the rural areas), it appears that forms of energy that require development in the future must be those to which the rural population of the country has easy access. In any case, a large amount of forest resources are currently being consumed in wasteful and inefficient ways throughout the country. It is tragic that in spite of the major social cost of depletion of forests in the country, official agencies and society in general have not addressed themselves to the problem of efficient utilization of forest resources. It is essential that research and development efforts be made toward not only the problem of growing fast-growing forests, particularly in those areas where large industrialization has taken place, but also for using forest products in more efficient ways for energy production. Some of the alternatives mentioned above can be extended to a number of rural areas, which once available, are likely to lead to more efficient utilization of fuel wood. In this context, it must be mentioned that a very clear and useful forest policy has to be evolved in the country in the near future. Since the benefits of energy are only a small part of the benefits from afforestation, it requires attention by other ministries involved in the field, apart from the Ministry of Energy itself. A great deal of damage has been done by lack of implementation of forest policy by state governments, where large-scale felling has been carried out as a result of favors to contractors who have been motivated mainly by the

TECHNOLOGICAL ALTERNATIVES 143

desire to "make a fast buck" at the cost of society. The system of award of contracts for cutting forests needs complete overhaul; it has often been a major source of corruption.

ORGANIZATIONAL ASPECTS

In the previous pages, we have discussed a large number of technological alternatives and mentioned some of the work that has already been done in the field to achieve more efficient and widespread use of the energy sources available in India. For a developing country such as India, investments in technology, with the high degree of uncertainty associated with such activities, make choices and priorities difficult to arrive at. Whereas the field covered by possible technological alternatives in the energy picture is vast, choices have to be very carefully exercised in allocating resources for further research and development. This can only be done on the basis of a detailed study of the state of the art, the resources required for further development, and the likely payoff from such developments. The FPC went into this question in evolving a technology plan for the energy sector.[8] In their recommendations, they mentioned certain priority areas. Most important among these were the development of the FBR, development of boiler designs to reduce oil consumption for thermal power generation, development of fluidized bed technology and a commercial generating plant based on this technology, development of SNG production and transport technologies suitable for Indian conditions, and finally, development of technologies for manufacture of cheap smokeless fuels for use in residential consumption.

The organizational problems involved in administering a large technology plan are also of great importance. For instance, right up to the early 1970s, India did not make coordinated efforts in developing energy resources, and the realization that a broad energy policy is essential came only in the wake of the increase of petroleum prices in 1973/74. Various recommendations by committees and study groups have been made on how to coordinate research and development efforts in the future. The idea of an Institute of Energy Studies has been mooted by various groups, but this has not yet been established. It appears that based purely on the size of the problem and the difficulties in bringing about effective coordination, as well as to avoid unnecessary duplication, a large institution to coordinate technological activities in the field of energy studies is essential. Even though the establishment of a full-fledged Ministry of Energy has led to a great advance in policy making in this sphere, it is not possible for a wing of the government, such as the Ministry of Energy, to coordinate all the different aspects of research and development work; these must be coordinated by an appropriate institution in order to forge ahead in the field of energy studies. If efficient use is to be made of resources that are currently being expended in different institutions all over the country engaged in energy-related research, a coordinating agency, such as would be represented by an Institute of Energy Studies, is essential, and further delay in this establishment will only be detrimental to the progress of technology in the vital field of energy research and development. This does not mean that the functions of the Ministry of Energy would be partially taken over by the Institute of Energy Studies; it would only mean that some functions that are currently not being performed by the Ministry of Energy, and, therefore, resulting in suboptimal utilization of research and development efforts and resources, will be created. The Institute of Energy Studies, if established, can function in consonance with policies enunciated and administered by the Ministry of Energy. A complete reorganization of the work of research and development in the energy sector is essential if India is to move toward self-sufficiency and efficient utilization of its energy resources in the 1980s.

NOTES

1. These are discussed in detail in R. K. Pachauri, The Dynamics of Electrical Energy Supply and Demand: An Economic Analysis (New York: Praeger, 1975), pp. 84–87.

2. Ibid., pp. 101-2.

3. Kirit Parikh, Second India Studies: Energy (New Delhi: Macmillan Company of India, 1976), p. 122.

4. Report of the Fuel Policy Committee (New Delhi: Controller of Publications, 1974), p. 107.

5. Amulya K. N. Reddy and K. Krishna Prasad, "Technological Alternatives and the Indian Energy Crisis" (Paper delivered at the Seminar on Energy, Hyderabad, March 1976), pp. 62-63.

6. Arjun Makhijani, Energy and Agriculture in the Third World (Cambridge, Mass.: Ballinger, 1975), pp. 111-16.

7. S. S. Penner and L. Icerman, Energy, vol. 2: Non-Nuclear Technologies (Reading, Mass.: Addison-Wesley, 1975), p. 428.

8. Report of the Fuel Policy Committee, pp. 105-13.

CHAPTER

6

POWER SECTOR

The unique position of power in economic development justifies a study in depth of this sector. In fact, the index of power consumption per capita is often considered as being a reflection of the level of economic maturity of a society. Soon after independence, the Indian government embarked on a large-scale plan for expansion of electric power generation. The success of this endeavor can be seen in the figures shown in Table 6.1. Table 6.2 shows the rise in per capita consumption of electricity since the year 1960/61.

The FPC went into depth on the subject of power generation and its effect on economic growth.[1] Significantly, the average annual rate of growth in power generation during 1953/54–1960/61 was 10.7 percent; from 1960/61–1965/66, it was 12.85 percent; and from 1965/66–1971/72, it was 10.35 percent. In the period 1971/72–1973/74, the rate of growth slumped to approximately 5 percent annually. During the Fifth Five-Year Plan, as well as during the Sixth Five-Year Plan period, that is, up to 1983/84, it is expected that the rate will increase to 10.7 percent in terms of the forecasts of the FPC. The forecasts are based on the assumption that all urban households and 70 percent of rural households in the country will be provided with electricity for lighting and the number of pump sets for agricultural purposes will increase from 2.5 million in 1973/74 to approximately 12 million in 1990/91. The installed capacity for power generation at the end of the fourth plan was 18,456 megawatts. In the first two years of the fifth plan, 3,524 megawatts were added, and it is expected that during the entire fifth plan, 12,500 megawatts will be added. At the end of the plan, there would be projects still in hand that would contribute 6,000 megawatts as spillover into the sixth plan period. The various expenditures that are proposed for the fifth plan and the different purposes for which they would be made are shown in Table

TABLE 6.1

Electricity Generation, 1960/61-1990/91

Year	Millions of Kilowatt Hours
1960/61	20,123
1965/66	36,825
1971/72	66,384
1978/79*	124,000
1983/84*	205,000
1990/91*	392,000

*Projected.

Source: Report of the Fuel Policy Committee (New Delhi: Controller of Publications, 1974), Table 9.1, p. 74.

6.3. The breakup by region of installed capacity as expected at the end of the fifth plan is shown in Table 6.4.

One of the factors that has inhibited rational planning of power in the country has been an adherence to outdated methods for forecasting power and energy demand. The power sector in the country is largely administered by state electricity boards, each exercising

TABLE 6.2

Per Capita Consumption of Electricity, 1960/61-1990/91

Year	Per Capita Consumption (kwh)
1960/61	38.20
1965/66	61.83
1970/71	89.76
1973/74	103.22
1978/79	173.40
1983/84	253.00
1990/91	447.00

Source: Report of the Fuel Policy Committee (New Delhi: Controller of Publications, 1976), Table 9.3, p. 74.

TABLE 6.3

Fifth Five-Year Plan: Power—Financial Outlay
(rupees in crores)

Item	States	Union Territories	Center	Total	Draft Plan
Generation	3,722.71	6.52	665.24	4,394.47	3,323.81
Transmission and distribution	1,897.73	78.78	104.74	2,081.25	1,634.27
Rural electrification MNP and state plan	360.54	10.74	—	371.28	698.24
Rural Electrification Corporation	314.02	—	—	314.02	400.00
Survey and investigations	74.92	2.72	55.24	132.88	138.68
Total	6,369.92	98.76	825.22	7,293.90	6,190.00

Source: Government of India, Planning Commission, Fifth Five Year Plan, 1974–79 (New Delhi: Controller of Publications, 1976), p. 60.

its jurisdiction over the different states of the country. Planning at the level of state electricity boards has been highly tentative and ad hoc. Since major power generation schemes are sanctioned and financed by the central government, forecasting and planning have been in the hands of the central government. Recent legislation has given the CEA wide powers in planning and administration of the entire power sector of the country.

Mention must be made of the forecasting methodologies followed by the central government. The CEA uses three techniques in projecting long-term demand for electrical energy. These are the trend method, the end-use method, and Scheer's formula.

In the trend method, the modified exponential equation of $Y = 3411.39 + 8555.05\, e^{0.0988X}$ (where Y equals energy requirement in millions of kilowatt hours; X equals time in years; and e equals Naperian constant) is used as it is found to be most suitable under Indian conditions by giving the best fit with past data. The results are obtained with the help of a computer.

In the end-use method, the energy consumption of various categories of load are estimated and added up to arrive at the total consumption. In the case of industrial consumption, the targeted produc-

TABLE 6.4

Breakdown of the Installed Capacity at the End of the Fourth Five-Year Plan and Fifth Five-Year Plan, by Region, by Type of Plant, as of March 31, 1974 and March 31, 1979
(megawatts)

Region	As of March 31, 1974				As of March 31, 1979			
	Hydro-electric	Thermal	Nuclear	Total	Hydro-electric	Thermal	Nuclear	Total
Northern	2,200	1,759	220	4,179	4,005	4,379	440	8,824
Western	1,037	2,612	420	4,069	1,760	5,042	420	7,222
Southern	3,080	1,437	—	4,517	4,738	2,387	235	7,360
Eastern	580	3,102	—	3,682	977	4,462	—	5,439
Northeastern	67	147	—	214	138	177	—	315
Other Union Territories	—	3	—	3	—	3	—	3
Total utilities	6,964	9,060	640	16,664	11,618	16,450	1,095	29,163
Total nonutilities	—	—	—	1,792	—	—	—	1,792
Grand total	—	—	—	18,456	—	—	—	30,955

Source: Government of India, Planning Commission, Fifth Five Year Plan, 1974–79 (New Delhi: Controller of Publications, 1976), Annexure 29, p. 130.

tion of the core and major industries of the country, as envisaged by the Planning Commission, are taken into account.

Scheer's formula is based on the thesis that for every 100-fold increase in the per capita generation, the rate of growth in generation will be reduced by half. Scheer developed the following formula after studying the load growth of a number of countries in the world, including India:

$$g = \frac{10^c}{u^{0.15}}$$

where g is the annual percentage growth in generation, u the per capita generation, and c a constant, which is 0.02 times population growth rate plus 1.330. The forecast of generation requirement is made with the help of this formula on a year to year basis with the estimated figures of population.

The <u>Ninth Annual Power Survey</u> of the CEA (published in 1975) has arrived at the forecasts of demand shown in Table 3.5. In addition, we have shown in Table 6.5 the forecasts obtained by each of the above three methods. This author has carried out a detailed study of the demand for electrical energy in the state of Andhra Pradesh. The attempt in the study was to estimate a number of econometric models that would represent the relationship between demand for power in different sectors with a number of economic-demographic variables. Some of the results are provided in Tables 6.6, 6.7, 6.8, and 6.9.

The models representing specific sectors within the Andhra Pradesh region were specified and estimated in two parts. The first related to relationships between number of customers of electricity in each sector or electricity rate group, and the second to average consumption in kilowatt hours per customer in each of these sectors. Hence, the dependent variables regressed were number of customers and kilowatt hours per customer respectively in each case against a number of alternative sets of independent variables. The models were estimated both in the simple linear and log linear forms, and the final versions presented in Tables 6.6 through 6.9 are those selected from a large number actually estimated on the basis of better fit and underlying economic rationale.

The lack of significance in some of the models estimated by us is due to high-degree multicollinearity in the data set and the lack of specific data at the state level for a number of variables (which might have helped increase the explanatory power of each of the models estimated). Using the results of the study, simulations were carried out to project demand up to the year 1990/91, based on a set of reasonable assumptions regarding growth parameters for the future.

This yielded results significantly lower than those of the forecast for the state by the CEA in its Ninth Annual Power Survey. This indicates that the forecasts of the CEA are likely to be on the higher side, which may be due to the predominance of past trends in the methodologies followed by them. Even in the end-use method, which breaks down demand for power into its component end uses, certain norms are used and projected forward, such as, say, power consumed per metric ton of aluminum produced. These norms do not take into account the effect of prices and technological change and, therefore, are likely to give unreliable forecasts.

The practice of using outdated forecasting methods is not peculiar to developing nations like India. In fact, the data base and methods being used for planning in India are surprisingly advanced if seen in relation to some other nations of the world. This author, in a previous book based on data for a region of the United States, pointed out some of the flaws and inadequacies of forecasting methodologies used by American utilities.[2] That situation, too, is undergoing change due to the need for long-range reliable forecasts forced on utilities by the long gestation period of nuclear and larger-sized thermal plants.

MANAGERIAL AND ADMINISTRATIVE PROBLEMS IN THE POWER SECTOR

The rapid growth of power generation in the country has naturally brought about severe managerial and administrative problems. These have resulted as much from the nature of the tasks involved as from

TABLE 6.5

Forecasts of Demand for Electrical Energy by Three Alternative Methods
(billions of kilowatt hours)

Method	Energy Demand	
	1983/84	1990/91
Trend (modified exponential form)	220.00	442.7
Scheer's formula	214.50	393.3
End use	211.30	385.0

Source: Government of India, Central Electricity Authority, Ninth Annual Power Survey (New Delhi, 1975).

TABLE 6.6

Estimated Models for Electricity Demand in the Domestic Sector of Andhra Pradesh

a) Customer Model
Form: Simple linear R^2: 0.996245 $\bar{R}^2$: 0.994368
Dependent Variable: Number of customers

	Independent Variables		
	Number of Villages Electrified	Urban Population	Constant
Unit	—	—	—
Regression coefficients	54.379639	0.094028	-529,743.190000
Standard deviation	7.981791	0.026366	157,639.370000
T-ratio	6.8130000	3.5660000	-3.3604746

b) Unit Consumption per Customer Model
Form: Log linear R^2: 0.321621 $\bar{R}^2$: 0.287272
Dependent Variable: Kilowatt hours per customer

	Independent Variables				
	Kilowatt Hours per Customer (t-1)	Average Price of Electricity (t-1)	Per Capita Income (t-1)	Time Trend	Constant
Unit	—	Paisas	Paisas	—	—
Regression coefficient	0.520173	0.126728	0.191151	0.019741	0.197999
Standard deviation	0.0942120	0.0870290	0.2247660	0.0149330	0.3870908
T-ratio	5.52100000	1.45600000	0.85000000	1.32220000	0.51068956

Note: One paisa equals 1 percent of a rupee; (t-1) relates to values pertaining to the previous time period.

Source: Compiled by the author.

TABLE 6.7

Estimated Models for Electricity Demand in the Commercial Sector of Andhra Pradesh

a) Customer Model
Form: Log linear R^2: 0.910021 $\bar{R}^2$: 0.865052
Dependent Variable: Number of customers

Unit	Independent Variables		
	Total Income	Urban Population	Constant
	Rupees in lakhs		—
Regression coefficients	0.6097120	1.4763950	-8.0028143
Standard deviation	0.5610640	0.8292600	3.4687896
T-ratio	1.0970000	1.7800000	-2.3476992

b) Unit Consumption per Customer Model
Form: Log linear R^2: 0.234572 $\bar{R}^2$: 0.203542
Dependent Variables: Kilowatt hours per customer

Unit	Independent Variables			
	Kilowatt Hours per Customer (t-1)	Population (t-1)	Time Trend	Constant
	—	1,000	1/100	—
Regression coefficients	0.4288130	0.6940960	-0.0357070	-1.4098701
Standard deviation	0.1082180	0.4107420	0.0203900	1.4174023
T-ratio	3.9650000	1.6900000	1.7510000	-0.99468595

Note: The (t-1) relates to values pertaining to the previous time period.

Source: Compiled by the author.

TABLE 6.8

Estimated Models for Electricity Demand in the Industrial High Tension Sector of Andhra Pradesh

a) Customer Model
Form: Log linear R^2: 0.994153 $\bar{R}^2$: 0.988340
Dependent Variables: Number of customers (1/100)

	Independent Variables	
	Number of Customers (t-1)	Constant
Unit	1/100	—
Regression coefficients	1.0170350	-0.0290308
Standard deviation	0.06377800	0.30535567
T-ratio	15.9460000	-0.0950731

b) Unit Consumption per Customer Model
Form: Log linear R^2: 0.510863 $\bar{R}^2$: 0.491682
Dependent Variable: Kilowatt hours per customer

	Independent Variables		
	Kilowatt Hours per Customer (t-1)	Time Trend	Constant
Unit	—	1/1000	—
Regression coefficients	0.5518860	-0.0358990	2.4776564
Standard deviation	0.11379400	0.01581100	0.62690687
T-ratio	4.8500000	2.2700000	3.9521914

Note: (t-1) relates to values pertaining to the previous time period.

Source: Compiled by the author.

TABLE 6.9

Estimated Models for Electricity Demand in the Agricultural Sector (Low Tension and High Tension) of Andhra Pradesh

a) **Customer Model**
Form: Log linear R^2: 0.993079 $\bar{R}^2$: 0.99031
Dependent Variable: Number of customers

	Independent Variables		
	Number of Villages Electrified	Average Price of Electricity	Constant
Unit	—	1/100 paisa/unit	—
Regression coefficients	1.2670360	−0.2362380	1.2401323
Standard deviation	0.05719200	0.08958100	0.49808174
T-ratio	22.1540000	2.6370000	2.4898167

b) **Unit Consumption per Customer Model**
Form: Log linear R^2: 0.527813 $\bar{R}^2$: 0.503904
Dependent Variable: Kilowatt hours per customer

	Independent Variables				
	Kilowatt Hours per Customer (t-1)	Seasonal Factor	Average Rainfall	Time Trend	Constant
Unit	—	1/100	Millimeters		—
Regression coefficient	0.4520860	1.3659770	−0.1391740	−0.3439610	7.6417236
Standard deviation	0.084616	0.319252	0.050066	0.156420	34.741150
T-ratio	5.3430000	4.2790000	2.7800000	2.1990000	0.2199617

Note: One paisa equals 1 percent of a rupee; (t-1) relates to values pertaining to the previous time period.

Source: Compiled by the author.

the lack of trained personnel and institutional weaknesses in the organization of the state electricity boards. The state electricity boards function directly under the state governments and are, therefore, subject to various types of political pressures, which often dilute their accountability in terms of financial and operating results. Utilities all over the world have recently undergone a series of financial strains brought about by high inflation, downturn in economic growth, and increases in fuel prices. In the case of India, these factors have been compounded by a lack of managerial and organizational strength and poor management practices.

Pricing of power by the state electricity boards has been unsound. The revenues of the state electricity boards have, therefore, been totally inadequate, not only for meeting the interest charges to government but also for basic requirements, such as proper maintenance. The electricity boards have been beneficiaries of low-interest long-term loans, which are not available to a large number of private sector organizations engaged in other spheres of activity. Interest rates charged by government for loans to electricity boards have been as low as 5.25, 5.50, 6.50, 6.75, 7, and 8 percent for some years in the period 1964/65–1973/74. In 1974/75, interest rates were raised to 11 percent. The plight of electricity boards can be seen from the arrears of interest that were outstanding at the end of 1973/74, as reported by the Finance Commission (shown in Table 6.10).

State governments are generally lax in the enforcement of loan obligations. This leads to permissive standards of performance on the part of the electricity boards. In a recent meeting of the National Development Council, the deputy chairman of the Planning Commission lamented the fact that electricity boards and irrigation authorities were responsible for a deficit of a total of Rs 300 crores for the whole country in one year. This has been brought about as much by lack of financial discipline as by improper pricing and lack of accountability. Whereas electricity boards have been the beneficiaries of many favors in the past, they are often saddled with so-called social obligations, which they have to accept due to their subservience to state governments. In this context, it is interesting to read the resolution on power and irrigation systems of the National Development Council, which is the top planning body in the country.[3] The text of this resolution is reproduced below:

> Heavy investments have been made by the country in irrigation and power systems and it is certain that these sectors will, in the foreseeable future, continue to absorb a large share of plan resources. It is, therefore, a matter of prime importance that these sectors should no longer be a burden on the State's finances but should contribute something to them.

TABLE 6.10

Arrears of Interest Outstanding from State Electricity Boards on Loans from State Governments, as of End of 1973/74

State	Arrears (crores of rupees)*
Bihar	81.08
Punjab	49.30
Rajasthan	42.00
West Bengal	35.02
Assam	34.71
Karnataka	32.66
Kerala	25.26
Haryana	24.59
Gujarat	24.59
Andhra Pradesh	18.09
Madhya Pradesh	12.06
Maharashtra	9.00
Jammu and Kashmir	6.46
Himachal Pradesh	4.19
Orissa	4.15

*One crore equals 10 million.

Source: Report of the Finance Commission (New Delhi: Controller of Publications, 1974), p. 208.

In the case of power systems, while there is scope to raise tariffs, the improvement in financial results should come largely from a higher level of utilization of the existing capacity, particularly thermal power plants, reduction in overheads and operating expenses, reduction in transmission and distribution losses, better collection of dues, prevention of theft and timely completion of projects. In addition, full advantage should be taken of the opportunities for the exchange of surplus power between States and regions and for integrated operation of hydel and thermal plants so that capacities are optimally utilized.

The major opportunity for covering expenses and increasing revenues in irrigation projects lies in raising water rates whenever they are far below the cost. There is also considerable scope for improved management of the irrigation systems and for ensuring that projects are implemented on schedule.

Taking note of the above—

The National Development Council hereby resolves that irrigation systems should cover working expenses and yield, if possible, something more and that power systems should cover working expenses and yield reasonable returns on investment by taking steps expeditiously to

(1) make maximum use of the capacity already created in the power and irrigation systems,
(2) reduce costs by cutting overheads and operating expenses, minimising losses and thefts and improving collection of dues,
(3) complete projects on schedule through efficient project management,
(4) raise rates where necessary.

One of the major grouses of the state electricity boards is having to implement schemes for rural electrification. These schemes in financial terms are unremunerative and often result in increase of the losses suffered by electricity boards.

The Rural Electrification Corporation (REC) was established in July 1969 as a result of the recommendations of the All India Rural Group Credit Review Committee. This committee came to the conclusion that for the development of agriculture in the country, an ambitious program of rural electrification was a must. This course of action became essential due to the green revolution, which relied largely on multicropping and year-round irrigation, much of which had to be provided by lifting groundwater with the use of electrical power. It was recommended that rural electrification, which had made sporadic progress in different parts of the country, be tackled on a national basis. One of the considerations that was raised at the time of the establishment of the REC was that while agricultural development would benefit through the efforts of this corporation, it must be ensured that the state electricity boards did not suffer any financial difficulties as a consequence. At that stage, it was also laid down that rural electrification schemes in the Fourth Five-Year Plan should be project based, development oriented, and accepted only on the basis of sound financial criteria. The REC, it was felt, should sanction schemes and monitor them on the basis of well-defined financial criteria. The charter of the REC laid down the following objectives:

To finance rural electrification schemes in the country
To subscribe to special rural electrification bonds that

may be issued by the state electricity boards on conditions to be stipulated from time to time

To promote and finance rural electric cooperatives in the country

To administer the moneys received from time to time from the Government of India and other sources as grants or otherwise for the purposes of financing rural electrification in the country in general.

In carrying out the above objective, the REC is guided by the directives given to it by the government of India, as provided for under Article 126 of the Articles of Association. These directives require the REC to:

1. establish sound policies and procedures for consideration, approval and implementation of rural electrification schemes to be financed by it;
2. develop and apply criteria for establishing priorities as regards the choice of schemes and the basis of economic viability; and
3. adopt a project approach so that extension of electricity along with investments result in increased agricultural production in the area.

The above indicates that the functions of the REC extend beyond merely providing financial assistance to the state electricity boards for expanding their rural distribution systems. Its basic responsibility is to ensure that the schemes it finances are established in accordance with sound norms of economic viability and as part of coordinated efforts in different areas to promote overall rural development and increase in agricultural production.

The performance of the REC, however, has been mixed. The projects it helps state electricity boards to initiate are not supposed to saddle the particular board with losses. It is supposed to finance projects that generate adequate returns to pay for the costs and to provide reasonable returns on the investments made. Obviously, this is possible only if the users of such schemes are able to increase their earnings substantially to pay for the use of electricity. Unfortunately, in a number of cases, this is not happening, and under pressure, state electricity boards have had to sell power to rural consumers at subsidized rates, thereby eroding their own total revenues and passing on this burden, in some cases, to other consumers.

The REC is supposed to go into all aspects of a project before it is financed. It pays particular attention to estimates of costs and is supposed to satisfy itself that once electricity is available, adequate demand will exist to support the investments made in the area. It also assesses the availability of groundwater resources, the potential for development of small-scale industries, and other aspects of economic activity on the basis of which demand projections for power are made. The REC then works out the net return that must be ensured on the basis of estimates available and takes account of the period within which this can be achieved. The net return is arrived at after allowing for demand, interests on the loan, and depreciation, operation and maintenance charges, and other costs of generating power. For rural electrification to be considered financially viable, it is required to earn a net return of 3.5 percent within a specific period, which is generally 15 years for areas that are supposed to have certain advantages for electrification and somewhat longer for others.

The REC also concerns itself with monitoring the projects that it sanctions. To ensure that development in the area concerned takes place on an integrated basis, the time horizon for projects that are sanctioned extends in some places up to 25 years. The monitoring of projects itself is a fairly extensive task. For instance, during 1975/76, the REC had been consistently monitoring 1,144 schemes out of a total of 1,239 that had been sanctioned by the end of that fiscal year (March 1976). The monitoring often requires coordination with other local programs in agricultural and industrial production, as well as with the tribal development and rural employment schemes that are in existence under some programs of the state or the central government.

The functioning of the REC has certain inherent economies of scale. These economies can be substantial when considering the large scale of its operations. For instance, the experience gathered by this corporation has enabled it to pursue standardization in specifications for materials and practices. This is done in close coordination with the CEA, the Indian Standards Institution, and the state electricity boards. It also carries out research, development, and training activities, particularly with regard to projects that focus on rural development schemes. To provide impetus to these activities, the REC has set up a high-level committee for coordination of research, development, evaluation, and training.

Mention must be made of some of the schemes of rural electrification that complement the basic objective of financing new rural electrification projects by the REC. For instance, for states that are below the national average of rural electrification and for certain backward areas, where lack of transmission lines has been a constraint in the extension of rural electrification, a special category of loans, under the so-called ST scheme for financing such transmission systems, has been provided. Another loan scheme, the so-called SS scheme, seeks to finance investments in certain system improvement measures, such as strengthening the capacities of substations and conductors and providing for shunt capacitors with the objective of reducing the line losses wherever they are abnormally high. A reduction in line losses of almost 10 percent is anticipated as a direct result of the SS schemes sanctioned by the corporation.

The financial results of the REC are shown in Figure 6.1. Effective as of July 1, 1975, the REC became a public limited company, permitting it to issue bonds for sale to the public. The scale of operations of the REC can be visualized by studying its operations during the year 1975/76. During this year, the REC approved 284 new projects in a total of 21 states and provided a total loan assistance of Rs 111.85 crores, as against 340 new projects and loan sanctions of Rs 136.88 crores during the previous fiscal year. One reason for a drop in the commitment of new loans was because of a larger financial outlay in completion of schemes that were in ongoing and advanced stages of execution.

During the year 1975/76, the REC also entered into two new categories of loans. These were known as the miniindustrial estates loan and minipilot single-wire earth return system loan. The miniindustrial estates loan is basically intended to facilitate extension of electricity to industrial estates in rural areas. This type of assistance is available only to villages, small towns, and some urban areas with a population not exceeding 20,000. These loans are for ten years and carry interest at the uniform rate of 7.75 percent per annum, with a moratorium on repayment of principal for the first two years. The loan outlay for individual projects is limited to Rs 15 crores only.

The minipilot single-wire earth return system loan was established to encourage adoption of single-wire earth return systems in rural areas, since it was found that in areas where the possibilities of development of electric load were not substantial in the near future, this system resulted in appreciable economies in the total project cost. Under this category of loan, the cost of an individual project would normally be between five to ten lakhs of rupees.* The period of loan is ten years, again with a moratorium on the payment of the principal for the first two years. Table 6.11 shows the physical achievements of REC schemes up to March 31, 1976. Each of the columns shown in this table has been broken down to indicate the target that had been laid down and the actual achievement.

The significance of rural electrification can be seen from a perspective view of agricultural development in the country. Of the total geographical area of 327 million hectares in the country, about 158 hectares of land is under cultivation. The average yield from Indian

---

*One lakh equals 100,000 rupees.

POWER SECTOR 161

FIGURE 6.1

Financial Results of the Rural Electrification Corporation

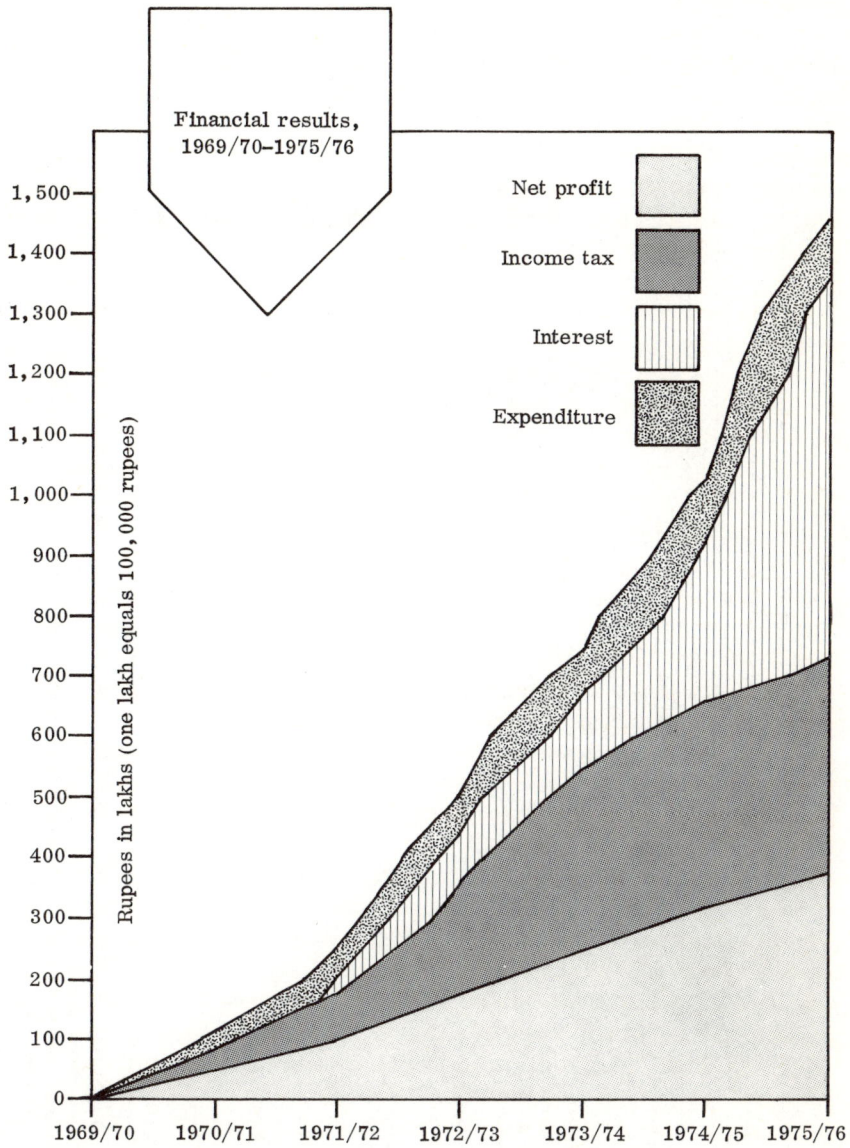

Source: Rural Electrification Corporation, 1975 Annual Report (New Delhi: Rural Electrification Corporation, 1976).

TABLE 6.11

Physical Achievements of Rural Electrification Corporation Schemes as of March 31, 1976[a]

State	Village Electrification[b]		Pump Sets		Small Industries	
	Phased Target[c]	Achievement[d]	Phased Target[c]	Achievement[d]	Phased Target[c]	Achievement[d]
Andhra Pradesh	1,249	1,143	25,632	16,854	1,518	696
Assam	220	157	107	32	573	141
Bihar	2,831	1,271	24,179	5,997	3,791	711
Gujarat	802	594	20,582	8,057	1,135	997
Haryana	90	90	14,803	8,252	1,692	942
Himachal Pradesh	1,379	967	427	176	627	386
Jammu and Kashmir	338	238	96	19	529	143
Karnataka	827	790	13,771	7,425	813	617
Kerala	150	103	4,826	1,946	777	385
Madhya Pradesh	1,912	1,293	31,084	11,740	2,396	1,036
Maharashtra	2,935	2,765	38,355	21,347	4,080	2,729
Manipur	—	—	—	—	—	—
Meghalaya	18	18	18	—	45	6
Nagaland	—	—	—	—	—	—
Orissa	1,750	1,906	12,838	844	1,567	242
Punjab	1,796	1,687	18,536	9,284	4,675	533
Rajasthan	1,636	1,568	25,802	13,072	2,671	1,363
Sikkim	—	—	—	—	—	—
Tamil Nadu	825	737	26,311	15,566	460	488
Tripura	—	—	—	—	—	—
Uttar Pradesh	3,360	2,113	19,191	9,458	7,372	331
West Bengal	3,991	3,622	11,168	3,815	6,540	1,153
Total	26,109	21,062	287,726	133,884	41,363	12,899

[a]Only includes area schemes that have been implemented for a period of three years or more.
[b]New villages only.
[c]Phased target in relation to the number of installments of loan drawn before September 1975.
[d]Achievement as of March 31, 1976.

Source: Rural Electrification Corporation, Seventh Annual Report, 1975-76 (New Delhi: 1976), pp. 50, 51.

POWER SECTOR 163

agriculture is extremely low due to poor irrigation and other inputs. The monsoon rainfall, on which a substantial part of the country relies for its irrigation needs, is unpredictable and varies in nature. There is enough empirical evidence to show that the uncertainties that are associated with surface irrigation do not affect groundwater resources in the same way. It is in this respect that rural electrification has to play an important role, by increasing the number of tube wells and pump sets in the country. Table 6.12 gives the groundwater resources estimated in the country. Current estimates of the extent of use of groundwater resources places this figure at below 40 percent of the total potential.

The achievement of rural electrification presented earlier can be seen in better perspective when compared with progress on electrification in rural areas, as shown in Table 6.13. It is obvious that rural electrification can have social benefits greatly in excess of private benefits to the direct users of this scheme. The availability of power in rural areas opens up the possibility for the implementation of various types of schemes, which would lead to overall development of the rural sector in the country. However, schemes that have been implemented have not generally been successful. This has been due as much to the lack of increase in the demand for power in rural areas covered by electrification as to the lack of interest by the state electricity boards, who have often suffered heavy losses because of these schemes, since they have been unable to recover the price of the power increase (in accordance with marginal costs of providing services to rural areas). A fresh look, therefore, at the achievement and failures of rural electrification and the organization of this vital activity is necessary to provide impetus in this area in the future.

It must be emphasized, however, that it is not merely the inability of electricity boards to charge rural consumers rational prices but also various problems of a technical nature that erode the earnings of state electricity boards. The line losses for the country as a whole are generally in the region of 20 percent, which compares extremely unfavorably with the standards observed in some other countries. By comparison, average line losses in West Germany are 5.9 percent; in Italy, 8.5 percent; in the Soviet Union, 7.5 percent; in the United States, 7.85 percent; in France, 7.10 percent; in Japan, 7.8 percent; and in the United Kingdom, 7.15 percent. It is generally accepted now that a large part of these losses are contributed by the extension of rural electrification and, in particular, the operation of pump sets for irrigation purposes. Table 6.14 shows the totals by state for line losses as of April 1972. Technical solutions to this problem can reduce the degree of losses, but the financial viability of such schemes is still not clear. For instance, it is estimated that provision of low-tension shunt capacitors for all the agricultural

TABLE 6.12

Groundwater Resources

State	Net Groundwater Recharge (million acre feet)	Annual Draft by the End of 1967/68 (million acre feet)	Net Groundwater Recharge Available for Future Groundwater Development	Area Irrigated by Groundwater at Present (millions of acres)[a]
Andhra Pradesh	17.2	3.570	13.60	1.400
Assam region, including Nagaland, Northeast Frontier Agency, and so forth	16.7	0.030	16.70	—
Bihar	21.9	2.350	19.50	1.200
Delhi	0.3	—	—	—
Gujarat	10.2	4.130	6.10	—
Haryana	3.5	0.750	2.70	—
Himachal Pradesh	0.9	n.a.[b]	—	—
Jammu and Kashmir	4.0	0.001	4.03	0.030
Kerala	5.4	0.004	5.40	0.016
Madhya Pradesh	26.7	4.220	22.50	1.000
Tamil Nadu and Pondicherry	11.5	3.470	8.00	2.300
Maharashtra	12.6	3.410	9.20	2.000
Mysore	10.0	1.030	9.00	0.750
Punjab	6.9	3.300	3.60	3.500
Orissa	16.0	0.150	15.80	0.200
Rajasthan	3.4	2.070	1.40	3.000
Uttar Pradesh	35.5	17.920	17.60	9.000
West Bengal	16.1	0.360	15.70	0.100
Total	218.8	46.738	170.83	24.496

[a]Distribution on the proportional rate of the total irrigated area.
[b]Not available.

Source: Report of the Fuel Policy Committee (New Delhi: Controller of Publications, 1974), Table 9.19, p. 86.

TABLE 6.13

Electrified Villages

Population Range*	Total	Number Electrified as of				
		March 31, 1951	March 31, 1961	March 31, 1968	March 31, 1969	March 31, 1970
Up to 499	351,653	522	3,986	10,265	19,934	26,222
500 to 999	119,086	611	4,306	9,787	17,226	21,775
1,000 to 1,999	65,377	843	5,918	11,567	18,128	22,504
2,000 to 1,999	26,565	825	5,458	9,441	12,913	15,948
5,000 to 9,999	3,421	197	1,319	1,963	2,397	2,638
10,000 and above	776	134	560	647	682	693
Total	566,878	3,132	21,547	43,670	71,280	89,780

*1971 census.

Note: The data for electrified villages of Punjab and part of Uttar Pradesh are based on the 1951 census; the data of electrified villages for Tamil Nadu up to March 31, 1961 are based on the 1951 census, and, therefore, population breakdown for all periods is also up to March 31, 1961; the population breakdown of electrified villages and other figures have been estimated wherever actual figures are not available.

Source: Report of the Fuel Policy Committee (New Delhi: Controller of Publications, 1974), Table 9.20, p. 86.

pump sets in the country at an average cost per unit of Rs 200 would bring down the losses observed in the rural low-tension transmission network, but the overall capital costs may be much too large to make this a financially viable scheme. Schemes of this nature could bring about more efficient utilization of assets in the power sector, and the REC must direct its attention to the evaluation of such schemes.

One reason why electricity boards in India have failed to reach standards of financial performance that would compare with some of the more advanced organizations in the country is because these boards have generally been fashioned on the existing bureaucratic structure at the state government level. Traditionally, as a general rule, government departments in the country have shied away from making financial decisions on sound business principles. Political and union difficulties have often compounded the problems faced by these boards, resulting in prolonged periods of heavy losses and weak management at all levels. All of this has resulted in huge losses being incurred year after year, and it is a major drain on the national treasury. The poor financial returns of electricity boards in different states can be seen from Table 6.15.

THE POWER EQUIPMENT INDUSTRY IN INDIA

With the scale of growth in the electrical energy sector witnessed in the past and plans for further expansion in the future, the establishment of a large and modern industry for production of power equipment becomes economically viable. India has an infrastructure that would support the functioning of this industry so that it would be competitive with other leading manufacturers of the world. In keeping with the philosophy of economic development that the Indian government pursued in the first two decades of independence, efforts at building up the power equipment industry in the country were made in the public sector. Until the early 1950s, the country was dependent on large-scale imports for the plants and equipment required in setting up power-generating capacity. There were a large number of units that were manufacturing items, such as transformers, small motors, conductors, wires, lamps, and so forth, required for feeding the growth of electric power, but there was hardly any attempt to set up manufacture of heavy equipment required in power generation.

In 1955, the government decided to set up a plant in the public sector for production of power generation equipment (at Bhopal in Central India). This factory went into partial production in November 1969; it passed through difficult years with poor production, but it has now grown to be a successful example of enterprise in the public sector. In the Third Five-Year Plan, studies undertaken for the de-

TABLE 6.14

Statement of Line Losses, Length of Lines, Number of Agricultural Pump Sets, as of April 1972

State	Line Losses, 1971/72 (percent)	Number of Pump Sets Connected	Total Units Sold, 1971/72 (megaunits)	Units Sold in Low Tension, 1971/72 (megaunits)
Andhra Pradesh	26.84	216,182	2,329.80	1,231.20
Assam	16.87	136	313.80	160.50
Bihar	15.72	74,023	1,469.16	333.30
Gujarat	16.04	79,463	2,044.27	819.66
Haryana	28.52	107,373	1,066.22	670.64
Himachal Pradesh	12.00	106	116.68	115.55
Jammu and Kashmir	18.00	165	187.00	187.00
Kerala	17.29	23,010	1,501.88	387.79
Madhya Pradesh	19.24	89,505	2,006.77	577.45
Maharashtra	12.66	235,195	2,973.81	1,194.80
Karnataka	15.86	149,921[a]	3,352.46	1,005.54
Orissa	4.75	596	1,740.31	191.21
Punjab	24.19	97,397	1,267.91	1,047.68
Rajasthan	23.62	49,296	1,065.40	490.16
Tamil Nadu	18.80	598,004[b]	4,834.18	2,445.85
Uttar Pradesh	26.40	147,537	3,909.04	1,670.03
West Bengal	13.39	1,557	1,098.69	231.06
Total		1,869,970	31,227.38	12,759.46

[a]217,610 pump sets as of December 31, 1975.
[b]702,403 pump sets as of January 31, 1975.

Source: Rural Electrification Corporation, unpublished data.

TABLE 6.15

Financial Returns of State Electricity Boards, 1971/72–1973/74
(crores of rupees)*

State	Revenue Receipts	Revenue Expenditure	Operating Surplus	Block Capital	Operating Surplus as Percentage of Block Capital
Andhra Pradesh					
1971/72	45.49	25.96	11.59	273.19	4.2
1972/73	47.40	26.34	12.00	308.96	3.9
1973/74	56.59	31.80	14.47	347.06	4.2
Assam					
1971/72	5.71	3.29	0.33	80.43	0.4
1972/73	6.31	3.40	0.47	87.75	0.5
1973/74	7.45	3.84	0.52	96.36	0.5
Bihar					
1971/72	21.69	21.69	3.13	204.03	1.5
1972/73	34.67	22.58	4.69	230.74	2.0
1973/74	37.37	23.39	5.14	263.24	2.0
Gujarat					
1971/72	38.73	24.78	8.45	194.90	4.3
1972/73	44.12	28.84	9.06	230.22	3.9
1973/74	52.88	34.06	11.07	270.13	4.1
Haryana					
1971/72	19.90	11.33	4.93	150.89	3.3
1972/73	24.31	15.40	4.98	177.55	2.8
1973/74	29.64	21.50	4.49	215.19	2.1
Karnataka					
1971/72	35.52	21.60	7.95	159.84	5.0
1972/73	31.50	20.53	6.87	177.50	3.9
1973/74	35.25	23.04	7.49	199.06	3.8
Kerala					
1971/72	19.30	9.92	5.81	179.99	3.2
1972/73	21.71	12.10	5.81	203.72	7.9
1973/74	24.88	13.45	7.08	235.31	3.0

Madhya Pradesh				
1971/72	35.25	15.91	13.32	213.40	6.2
1972/73	39.70	18.04	14.92	242.40	6.2
1973/74	42.41	18.56	16.19	274.68	5.9
Maharashtra				
1971/72	62.33	34.26	19.27	332.87	5.8
1972/73	70.84	42.68	19.10	385.95	4.9
1973/74	90.50	15.83	28.32	432.09	6.6
Orissa				
1971/72	13.34	8.00	2.49	114.00	2.2
1972/73	16.75	8.93	4.48	129.00	3.5
1973/74	21.29	11.71	5.65	151.00	3.7
Punjab				
1971/72	20.85	10.99	5.69	224.16	2.5
1972/73	21.00	12.52	4.14	262.96	1.6
1973/74	22.92	14.87	2.62	314.02	0.8
Rajasthan				
1971/72	23.82	13.31	6.08	184.66	3.3
1972/73	26.08	14.15	6.51	222.77	2.9
1973/74	34.73	21.85	6.96	255.27	2.7
Tamilnadu				
1971/72	69.40	42.15	15.73	440.27	3.6
1972/73	65.10	42.25	7.26	481.31	3.8
1973/74	91.51	58.42	18.42	520.08	3.1
Uttar Pradesh				
1971/72	67.44	38.13	17.16	574.06	3.0
1972/73	81.45	43.01	24.86	648.59	3.8
1973/74	96.70	54.88	24.59	784.42	3.1
West Bengal				
1971/72	23.27	15.56	3.38	120.71	2.8
1972/73	24.72	17.85	3.87	137.86	2.8
1973/74	27.39	19.68	3.91	164.56	2.1

*One crore equals 10 million.

Source: H. V. Dayal, "State Electricity Boards: Growth without Profit," The Economic Times, September 29, 1976, p. 5.

mand for power generation equipment indicated that the factory at Bhopal would not be adequate to meet the entire demand of the country. It was therefore decided to set up other units at Hyderabad, Tiruchirapalli, and Hardwar. Some preliminary work for setting up these units was done at Bhopal itself, but in 1964, another corporation was set up in the public sector—Bharat Heavy Electricals Limited (BHEL)— which took over the management and control of these units from Heavy Electricals India at Bhopal. Of the new units, the first one to go into production was the high-pressure boiler plant at Tiruchirapalli in May 1965. This was soon followed by the heavy power equipment plant at Hyderabad in December 1965, and the heavy electrical equipment plant at Hardwar in January 1967. The unit at Bhopal continued to function as an independent company until January 1974, when all the four units were merged to form BHEL.

BHEL has emerged as an industrial giant and has a fairly well-established long-range corporate plan, which is carefully dovetailed with the five-year plans of the government. In the initial stages, considerable efforts had to be made to strengthen the engineering function in this corporation and to rationalize the manufacture of the products on a corporate basis in order to avoid duplication of work at its component units and to ensure optimum utilization of assets. Also, in its initial stages, BHEL provided poor after-sales service, but this has improved considerably in recent years.

Since BHEL was entrusted with the task of developing the entire heavy electrical equipment manufacturing program in the country, it expanded operations to the manufacture of power-generating equipment for thermal (including boiler), hydroelectric, and nuclear power stations. Additionally, it produced equipment for transmission and distribution of power, such as switch gears, transformers, and large-size motors. In recent years, it has also diversified its production to produce compressors, high-speed turbines, boiler auxiliaries, and so forth. In the last ten years, all the units of BHEL together have manufactured equipment worth Rs 15,000 crores. Its contributions have taken place not only in the power sector but in industries like sugar, mining, electrochemical, metallurgy, fertilizers, refineries, petrochemicals, and steel. For the electrification and dieselization program of the Indian Railways, BHEL has been producing traction motors to be fitted on to both the diesel and electric locomotives that are being produced in the production units of the Indian Railways. Recently, BHEL has also entered into the manufacture of onshore oil drilling rigs, which will be supplied to the major oil exploration and production agency in India, namely, the ONGC. This new item of production will not require any further investment in equipment, since the existing capacity in the company will be utilized for the purpose. The expertise developed in manufacturing some of the major items being produced by BHEL can be assessed from the developments required in manufacturing these items.

Steam turbines and their auxiliaries are being manufactured at Bhopal, Hardwar, and Hyderabad. Bhopal has manufactured turbosets of 30 and 129 megawatts and turbines of 236-megawatt rating. Hyderabad has manufactured 60- and 110-megawatt sets, and Hardwar has manufactured 100- and 200-megawatt sets. Manufacture of these items at Hardwar and Bhopal commenced in 1968/69 and at Hyderabad in 1967/68. These plants of BHEL have also developed their own expertise to deal with different sizes and types of steam turbosets for both the utility industry and other industrial applications. A fair degree of coordination takes place between the various units to assimilate the benefits of experience generated in each plant. Efforts are currently in hand to develop manufacture of 500-megawatt thermal sets by the early 1980s. The current emphasis by the government is on setting up superthermal power stations in the country of 1,500- and 2,000-megawatt capacity, for which BHEL is also planning to establish manufacture of large unit-sized (200- and 500-megawatt) sets. For the requirement of pumps for thermal plants, BHEL has developed designs for the manufacture of vertical as well as horizontal types for power generation and other applications. These pumps include boiler feed pumps, condensate pumps, circulating water pumps, and various types of oil pumps. Other manufacturers, such as Flowmore and Bharat Pumps and Compressors in the public sector are manufacturing centrifugal and recirculating pumps for the fertilizer, steel, and petrochemical industries.

Manufacture of boilers in the country has been taking place over a long period of time, but there are basically two manufacturers in the country that can produce boilers for power generation, namely, the Tiruchirapalli unit of BHEL and ACC-Vickers Babcock at Durgapur. Technical collaboration in these two units is with Combustion Engineering and Babcock-Wilcox of the United States. The product mix of the Tiruchirapalli unit covers a much wider range and includes manufacture of fans, blowers, electrostatic precipitators, valves, and so forth, which are not manufactured by ACC-Vickers Babcock.

Expansion plans at the moment are being undertaken in both of these factories (to manufacture boiler house components to a capacity of 2,500 mwe at the BHEL Tiruchirapalli unit and 1,000 mwe at ACC-Vickers Babcock). This capacity will be adequate to meet the domestic requirements of India for the next five to six years, as well as to meet the export programs that are envisaged over this period. In order to upgrade technology and bring about innovations in this field, BHEL has set up a National Boiler Heat and Mass Transfer Institute at Tiruchirapalli.

Equipment for hydroelectric stations is also being manufactured by BHEL at its units located in Bhopal and Hardwar. The Bhopal plant has been engaged in the design and manufacture of hydroelectric equipment since 1964/65 and the Hardwar plant since 1969/70. Nearly 5 million kilowatts of equipment for hydroelectric power stations will be delivered by the two plants during the period 1976/77-1979/80. The sizes of the hydroelectric sets manufactured in these two plants varies from 5 to 180 megawatts. With adequate design expertise, these plants can cater to different types of schemes, with Pelton-, Francis-, and Kaplan-type runners. They are also engaged in manufacturing and setting up some pump storage installations. With regard to the manufacture of hydroelectric sets, complete self-reliance has been achieved within the country.

There are 15 major manufacturers of power transformers in the country, of whom 5 are equipped to manufacture transformers of 220-kilovolt-ampere capacity. There are a number of manufacturers of current, potential, earthing, furnace, and mining transformers. Apart from BHEL, two other public sector units, namely, New Government Electric Factory at Bangalore and Transformers and Electricals at Kerala, are planning to manufacture power transformers of 400-kilovolt-ampere rating. Most other units manufacturing power transformers are engaged mainly in meeting the demand for transformers of up to 220-kilovolt-ampere rating. The production of transformers currently is at a level of about 12 million to 13 million kilovolt-amperes per annum.

There are also 15 major firms manufacturing circuit breakers of different ratings in the country. As far as 400-kilovolt-ampere circuit breakers are concerned, the only manufacturer is BHEL. The number of high-tension circuit breakers produced per annum in the country is around 7,000, with almost the same number for low-tension circuit breakers as well.

As mentioned earlier, for manufacture of electrical motors of small capacity, that is, below 200 horsepower, there are a large number of manufacturers in the country, and they are more than able to meet domestic demand, including some special performance electrical motors in this range. In fact, almost 70 percent of the total electrical motors produced in the country are in the range of up to 50-horsepower capacity. In the range including and above 200 horsepower, there are ten major manufacturers in the country. BHEL manufactures the largest range of motors, including direct current (DC) general purpose and rolling mill machines, DC mill/crane motors, alternating current (AC) induction motors (both squirrel cage and slip ring), AC synchronous motors, and AC alternators. In addition, special purpose motors, such as flameproof motors, geared motors, and traction motors for the Indian Railways, are also being manufactured within the country. There is, therefore, adequate capacity in the country for meeting the needs of various auxiliary motors that are required in power stations.

POWER SECTOR   173

   Manufacture of nuclear power plant equipment in India is being handled mainly by BHEL and Larsen & Toubro. For the future of the nuclear power program in the country, most equipment will be manufactured indigenously. An indication of the progress made in indigenous manufacture can be seen from the fact that whereas for the first unit of RAPP (India's first Candu-type nuclear power station), the foreign exchange spent on the project was 50 percent of the total, for the second unit of MAPP (the second Candu-type nuclear power station), this figure will be reduced to 20 percent, and for the Narora Atomic Power Project (NAPP) (the third Candu-type nuclear power station), total import content in terms of foreign exchange spent is not likely to exceed 15 percent. BHEL has already undertaken to supply equipment, including some turbosets, for both the units of MAPP from their Bhopal and Hardwar units, condensers and feed water heaters from their Bhopal unit, and some generators and heat exchangers from the Tiruchirapalli unit. Other components manufactured indigenously include calandras and end shields.

   The performance of BHEL in the past few years can be seen from Table 6.16. This indicates the progressive growth made by BHEL in recent years in accounting for a major share of the total power capacity installed in the country. The future plans for new installations in the country can be assessed from the company's order book (on March 31, 1976), as shown in Table 6.17. The total output of the company is shown in Table 6.18, which provides an indication of the growth in the range of its manufactures, as well as the total volume produced.

   Although, so far, there is cause for considerable satisfaction in the performance of BHEL (in fact, it is often referred to as a model of successful public enterprise in a developing society), BHEL will have to establish systems and develop expertise if it is to meet the challenges of the future. For instance, a greater emphasis on research and development is essential in the context of India's plans for expansion of its electric power systems. Investment in research and development of around 5 to 10 percent is typical of successful power equipment industries in the world. In the long run, it is imperative that a company in this field mount a major program in basic research and development in order to survive. Know-how can be acquired from other sources, but the indigenous industry cannot possibly grow if reliance is placed only on import of know-how.

   The past efforts of the Indian heavy electrical industry in managing and setting up large research and development efforts has been unsatisfactory. This is being changed through a major effort by BHEL. The research and development division of BHEL has been organized as a major corporate activity; a total of Rs 16 crores is being invested to equip and set up this division on a sound footing at Hy-

TABLE 6.16

Addition to Installed Capacity in the Country's Hydroelectric,
Thermal, and Nuclear Power Plants, 1971/72–1975/76
(megawatts)

Plant	1975/76	1974/75	1973/74	1972/73	1971/72
Thermal	890	1,170	320	500	400
Hydroelectric	914	550	179	300	200
Nuclear	—	—	—	180	—
Total	1,804	1,720	499	980	600
BHEL					
Thermal	770	930	320	90	150
Hydroelectric	619	426	26	15	30
Total	1,389	1,356	346	105	180

Source: Bharat Heavy Electricals, Annual Reports (New Delhi: Bharat Heavy Electricals Ltd., 1972, 1973, 1974, 1975, 1976).

derabad. A proposed complex for supporting the research and development division will include laboratories for investigation in basic sciences, for instance, materials sciences, insulation materials, aerodynamics, metallurgy, and so forth. A research institute is being set up at Tiruchirapalli, an ultrahigh voltage laboratory at Bhopal, and a national Boiler Heat and Mass Transfer Institute at Tiruchirapalli, as mentioned earlier.

With economies of scale being significant, the trend toward larger and larger unit sizes, particularly in thermal power stations, will continue. Therefore, the 500-megawatt set currently under development by BHEL will play an important part in the establishment of large thermal power stations in the future.

BHEL faces problems in supply of materials from Indian industry, particularly with respect to sophisticated castings and forgings, which require strong quality control and know-how in metallurgy. Even though significant strides have been made recently in developing materials, for instance, in insulating materials, steam conductors, and so forth, within the country, BHEL's own Central Foundry Forge Plant at Hardwar will have to meet a significant part of the demand for sophisticated castings and forgings. Similarly, the seamless steel tube plant that is being set up at Tiruchirapalli is another effort in vertical integration, which becomes important in developing countries

TABLE 6.17

BHEL's Order Position, as of March 31, 1976
(millions of rupees)

Generating Equipment	Orders Not Complied with at the End of the Year	Work in Progress	Finished Goods	Order Value Yet to Be Executed
Thermal and hydroelectric sets	4,900.0	340.0	560.0	4,000.0
Power utility boilers	2,290.0	150.0	270.0	1,870.0
Transmission and distribution equipment	1,030.0	160.0	150.0	720.0
Industrial and other products	2,880.0	270.0	460.0	2,150.0
Total	11,100.0	920.0	1,440.0	8,740.0*

*Value for manufacturing division alone was Rs 798 crores.

Source: Bharat Heavy Electricals, Annual Report for 1975-76 (New Delhi: Bharat Heavy Electricals, Ltd., 1976).

establishing manufacture of sophisticated products, which require advanced technologies to be employed at every stage of manufacture.

With the dominant role that BHEL is playing in the country, it becomes imperative that if advantage is to be taken of its traditions and know-how in moving into new fields of application, the research and development programs of the corporation must necessarily be geared toward establishing technologies that are likely to be developed on a commercial basis in the future. These would include moving into areas of application of solar and geothermal energy and magnetohydrodynamics. The Energy Systems and New Products Division of BHEL has been established for this purpose. Efforts are already in hand to evolve a suitable design of a solar-based ten-kilowatt generator in collaboration with the Indian Institute of Technology at Madras.

Some of the major items of research and development currently in hand at BHEL are mentioned below:

1. Design and manufacture of prototype fluidized bed boiler leading to commercial designs for industrial boilers.

TABLE 6.18

BHEL's Physical Output, Actuals, 1969/70–1975/76

Item	1969/70	1970/71	1971/72	1972/73	1973/74	1974/75	1975/76
Bhopal hydroelectric sets (megawatts)	96	75	15	314	609	595	755
Bhopal turbo sets (megawatts)	120	30	30	120	150	516	480
Hardwar hydroelectric sets (megawatts)	—	—	30	130	91	190	340
Hardwar turbo sets (megawatts)	100	100	100	300	700	1,080	1,250
Hyderabad turbo sets (megawatts)	180	230	100	390	550	780	390
Tiruchirapalli boilers (metric tons)	18,800	21,932	26,300	33,800	45,673	59,600	65,506
Bhopal							
Transformers (mva)	1,473	2,117	1,171	2,373	3,501	3,402	3,893
Switchgears (number)	1,307	1,869	1,758	1,909	1,959	2,268	2,547
Traction motors (number)	742	877	926	1,030	1,094	1,066	751
Industrial motors (number)	191	238	296	547	549	479	588
Hyderabad							
Switchgears (number)	139	194	401	514	577	774	851
Hardwar							
Industrial machines	225	226	75	22	233	324	360

Source: Bharat Heavy Electricals, Annual Report for 1975–76 (New Delhi: Bharat Heavy Electricals, Ltd., 1976).

2. Design of a steam turbine (16 megawatt) for the prototype FBTR for the Kalpakkam establishment of the DAE. This turbine has been taken up for manufacture by BHEL.

3. Development of conceptual designs for 500-megawatt boilers.

4. Development of a BHEL design of a high-pressure heater for 210-megawatt stations.

5. Design and development of prototype air-cooled condenser.

6. Development of runners for the 200-megawatt hydroelectric turbine for the power station being set up in the mountainous region of Tehri Garhwal.

7. Development of bulb-type straight flow hydroelectric turbine.

8. Setting up an ultrahigh voltage laboratory to support future developments.

BHEL also keeps in touch with developments in other countries in order to close the technology gap (by picking up know-how in specific areas—where the present Indian base does not justify development or in order to reduce the time or cost gap where necessary). One such basic need is in the area of large-size turbine generators. Other areas where similar action has been taken is in the field of industrial electronics, development of a combined cycle power system, and extra high voltage/ultrahigh voltage transmission systems and products. BHEL has also entered the export market, and during the year 1975/76, orders received were Rs 400 million in value, including an order worth Rs 170 million for three numbers of 120-megawatt boilers from the National Electricity Board in Malaysia. It is currently setting up a complete power station in Libya, which will be handed over to the Libyan government after it has been brought into operation.

It can be seen that the power sector in India has made significant strides. It is also easy to see that the challenges in the future will be far greater than those that have been met in the past. With the large-scale investments that have taken place in the manufacturing facilities for power equipment, the responsibility of Indian managers in this industry becomes much larger. The challenges of the new technology necessitate taking a long-range view, since building up know-how, capabilities, and skills takes a long period of time.

There is also an urgent need to streamline the administration of the electricity industry in the country. Whereas some improvement has taken place in the functioning of state electricity boards, the gaps between both the managerial skills and systems required to supply power sufficiently and efficiently in the future and the present level of capability available are likely to grow, unless a major reorganization effort is undertaken. Rural electrification obviously has an important role to play in the development of rural India. But priorities have to be reoriented to ensure that resources are not misallocated and optimal use is made of investments directed toward rural electrification. Unless the financial viability of state electricity boards (by a careful selection of rural electrification schemes) is ensured, the entire power program in the country will remain in a mess, and major dislocations in supply will continue to take place.

NOTES

1. Fuel Policy Committee, Report (New Delhi: Controller of Publications, 1975), pp. 74-75.
2. R. K. Pachauri, The Dynamics of Electrical Energy Supply and Demand: An Economic Analysis (New York: Praeger, 1975), pp. 3-5.
3. Government of India, Planning Commission, Fifth Five Year Plan, 1974-79 (New Delhi: Controller of Publications, 1976), p. 90.

CHAPTER 7
CONCLUSIONS

The foregoing pages have attempted to present the total picture of the energy sector of India. This effort would be incomplete without some attempt at integrating these various elements and arriving at some conclusions on which energy policy for the country should be based. Energy policy for a developing country must necessarily be viewed as a part of overall developmental policies and strategies. It would, therefore, be useful to spell out some of the objectives on which national programs must be evolved and implemented.

The concern for evolving a national energy policy in India arose out of the major increase in oil prices that took place in 1973-74. Quite naturally, the original shape that government decisions and plans took in the wake of these price increases were directed toward maintaining stability and equilibrium in the nation's economic activities. This concern, to a large extent, still dominates government plans and actions. The major emphasis in policy making is on minimizing economic disruption and dislocation as a result of constraints in supply of energy and in attempting to allocate energy resources in keeping with certain priorities that supposedly reflect national objectives.

Whereas this aspect of energy policy is justified in the short run, the long-term perspective of the problem needs to be brought into sharp focus. This will result only from deliberate planning to produce various types of energy at least cost of production, as well as of consumption. This would effectively mean that substitute forms of energy must be produced and supplied in the market in such a manner that the overall cost of substitution from existing forms is minimized.

Another element of policy making and planning in the energy sector that is of special relevance to developing countries deals with the role of energy as an essential prerequisite for social change and improvement in living standards, particularly in the rural sector. Past priorities and plans have often ignored the importance of energy supplies and consumption in rural India, which were perhaps based on an ignorance of the long-run benefits that would accrue to the country as a whole if cheap, efficient, and easily available sources of energy were to be introduced for consumption in the villages. Even where belated attempts have been made for extension of commercial sources of energy in rural areas, efforts made by the government and other agencies have been unproductive in terms of ensuring adequate social returns from the investments made. Schemes of investments controlled by the REC are significant examples of this.

The lack of effective and integrated energy policy until recent times is perhaps responsible for the absence of detailed facts and relevant data to undertake useful analyses and appraisals of the energy situation in the country. Even though the Planning Commission, as the central planning organization in the country, has been responsible for investments in new projects for energy generation and its consumption by different groups of consumers, there has been no coordinating agency, particularly for investments in new technologies, improvements in existing practices, and research and development efforts in the energy sector. The useful role that availability of information can play in bringing about optimal deployment and allocation of resources can hardly be overemphasized.

Various organizational lacunae have been responsible for the paucity of information and data for planning in the energy sector. The first of these can be identified as the lack of a central department or ministry to deal with problems of energy during the almost three decades of independent government. Of course, the absence of such an organizational focus has been present in almost every country of the world until the mid-1970s. The diffusion of plans and activities has resulted in considerable snags in coordination and integration of various elements in the energy sector. For instance, until quite recently, the central body for planning and monitoring of developments in the power generation industry was the Central Water and Power Commission. As a result of its very nature, problems of development in power generation were seen as being an adjunct to management of water resources in the country. Senior positions in this organization were naturally staffed by civil engineers, who in their own right may have been excellent water resource specialists but who had little familiarity with the complex problems of alternatives in power generation, economies of scale, and other aspects of power planning.

Another weakness, which unfortunately persists even today, is the lack of a central Institute of Energy Studies. A great deal of momentum has been observed in the efforts of various agencies and insti-

CONCLUSION

tutions in the country that are doing pioneering research work in energy-related problems, both in attempts to find "hardware" and "software" solutions, but it appears that the main wing of the government that deals with energy, namely, the Ministry of Energy, is neither equipped nor organized to deal with the problems of administering a coordinated research and development plan or in directing research on some of the policy issues related to the energy sector. Until a reorganization and clear definition of responsibilities is brought about at the top level of the government, it is unlikely that various organizations in the states or institutes dealing with energy can develop or implement a coherent and complete policy for the energy sector.

Of primary importance to a country that is not self-sufficient in energy is the development of new technologies. We have discussed some of the technological alternatives that are open to India today. With the scarce resources available in a developing society such as India, adoption of investment plans requires detailed analysis and evaluation.

In the short term, considerable improvement can be brought about by adopting vigorous policies for conservation and efficient utilization of energy. The concept of thermal efficiency would have to override any evaluation based on economic efficiency in this effort, since pricing in the energy sector does not truly reflect social costs of production and supply. As an example, one can see that 70 percent of the commercial energy in the country is consumed in mining, manufacturing, and transportation. Very little effort has been made either within these industries or by consultant organizations to suggest and adopt ways and means by which this amount of energy can be used more efficiently. A casual look at most trucks, buses, and so forth plying the roads of India would convince any observer that the fuel efficiencies of most vehicles based on their maintenance standards are deplorably poor.

Government has a vital role to play, by legislating against inefficient use and ensuring the reduction of waste in various forms of energy consumed today. This can perhaps be done by laying down minimum standards to ensure efficient use of energy in equipment and appliances. Testing and inspection organizations, such as the Indian Standards Institution, can ensure adoption of efficient designs and maintenance of minimum standards of quality in various appliances and equipment sold on the market. As an example, if electric pump sets are to be sold only as matched sets with built-in capacitors, the unusually large peak loads that are imposed by agricultural consumers on electric power utilities would be reduced significantly.

The government has made efforts to ensure that coal should be the primary source of commercial energy in this country. However, decades of developments based on falling real prices of petroleum and electricity and the various conveniences associated with using the latter forms of energy cannot possibly be reversed in a short period of time. The coal industry in the last few years has risen to the task of producing much greater tonnages of coal, but these efforts have not yielded adequate results due to the fact that forecasts of demand have not been realized. In the meantime, major investments have been made in the modernization and mechanization of coal mines, as well as in the handling and transportation facilities. These were based on the premise of large increases in demand for coal, which, if not realized, would result in large-scale losses to Coal India. Under these pressures, the Ministry of Energy is considering inviting public participation in the financial structure of Coal India. The slow movement of inventories and the accumulation of large stocks of coal at the pitheads of mines and stockyards of consumers have imposed an additional financial burden on the coal industry. The effect of large inventories and slower increase in demand has been to induce a slowdown in the growth of production.

It is safe to conclude that a major increase in demand for coal through substitution in existing uses and adoption of coal-based technologies can take place only with technological developments in the consumption of coal itself. Foremost among these are the technologies of coal gasification and production of smokeless varieties of coal for convenient domestic consumption. While investments in modernization of mining facilities are being made faster development of these technologies must also be considered, since probable payoff from investments in developing these will be of importance. These past years have been ones of neglect in developing coal gasification, and the government would do well to provide adequate financial and other support for a faster pace of development in the future.

Other areas in which new technologies deserve priority in development are in the use of solar energy and MHD. This author is not optimistic about gobar gas (biogas) plants making a major dent on the energy scene within the next five to ten years, mainly because of a lack of infrastructure and organizational facilities at the rural level, but also because of the uncertainties associated with economic benefits that are supposed to result from the operation of these plants. On the other hand, there appears to be a greater feasibility in developing gobar gas plants around urban and metropolitan areas, based on organized collection of dung and distribution of gas and fertilizers from large-scale plants of this type. Cooperative schemes extending into rural hinterlands of major cities have proved useful in such activities as milk collection, sale of vegetable produce, and so forth, and there is every reason to believe that large-sized gobar gas plants established around urban centers using similar arrangements could work efficiently.

CONCLUSION

There is currently some uncertainty in the growth of nuclear power generation in the country. Even though officially the fact is not admitted, there is no doubt that the slowdown in the development of nuclear power has been occasioned by political considerations. Contractual arrangements between India and the United States and India and Canada have recently run into rough weather. The first of these relates to the supply of enriched uranium to TAPP by the U.S. government. Even though the Indo-U.S. accord for the establishment of the Tarapur plant contains a clause that irrevocably commits the United States to supply adequate quantities of enriched uranium to run the station at optimum capacity during its entire lifetime, supplies have nevertheless been subjected to considerable controversy in the United States; during the latter part of 1976 and the early part of 1977, no supplies have been forthcoming. The U.S. government, under its new policy of discouraging proliferation of nuclear weapons, has been attempting to impose certain conditions, such as inspection of all nuclear power installations in the country, as a reprequisite for supply of enriched uranium to Tarapur. Even though President Jimmy Carter's new energy policy is flexible—in the sense that it permits him enough scope to exempt favored nations, such as India, from general controls on the supply of nuclear material—developments of the past few months have caused a certain amount of rethinking on the part of planners and policy makers in India. The virtual collapse of collaboration agreements between India and Canada in respect of the Rajasthan Atomic Power Plant and other nuclear programs as a consequence of India's nuclear blast of 1974 is another development that has caused rethinking in New Delhi.

Obviously, the nuclear power program in the country is not likely to take off again until the FBR becomes a commercial reality. Should this not happen in the near future, the social burdens of supporting a large nuclear program without any returns of scale of nuclear power projects would be expensive for a developing country. It is a little-known fact that nearly one-third of the country's scientific manpower in the governmental sector is engaged at the moment in nuclear research and development. Apart from the problems of not providing adequate professional challenges to these talented men and women, the costs of holding on to such a large manpower without major productive returns cannot possibly be supported in the long run.

At long last, India's oil and natural gas policy exhibits an enlightened approach. The major investments being made in the fifth plan period and the talented leadership at the top of the organization (for exploration and production) is adequate testimony to the government's vigorous approach. As mentioned previously, there is cause for reasonable optimism about increasing India's reserves of petroleum and natural gas and stepping up production in the near future.

The major failure in the petroleum sector has been, and to some extent continues to be, the absence of a rational pricing policy. Unless prices of petroleum products are increased progressively to reflect the true long-run social costs of providing this resource, consumption and exploitation will take place at a rapid pace, but it will be based on inefficient use in applications, whereas other substitutes could perhaps provide efficient alternatives. Major research work is warranted in determining the effect of prices on consumption of petroleum products and a proper evaluation of benefits from various types of uses. The development of the petrochemical industry is justified, and if need be, this industry should be supported by (differentially) low prices of petroleum-based inputs.

The problem of depletion of forest area in the country is another aspect that has been ignored in our developmental efforts. Our policy of afforestation has been based on random replacement of trees. The forest area under conservation is far below the optimum, as is indicated by the fact that the per capita area under forests in India is 0.14 hectares, as against the world average of 1.6 hectares. At the same time, the contribution of forestry to the GNP is negligible, totaling 1.2 percent in 1973/74, as against the total farm sector, which is close to 50 percent (even though forest area in the country is 50 percent of that under agriculture). Undoubtedly, the social returns from afforestation are much larger than private returns, and, therefore, the actual benefits to a country cannot easily be quantified. Nevertheless, a revised program to promote forests must be based on considerations of forestry as a farming activity, whereby quicker turnover of forests would add substantially to national income. It is unfortunate that this concept, even though adopted and pursued by other nations all over the world, has been neglected in our country.

A major direction in which energy policy must move is toward that of self-reliance. It is significant that even a country as richly endowed in energy wealth as the United States is currently making a major effort toward such self-reliance. For a country as vulnerable as India, the lessons of oil embargos, sudden price increases, revocation of nuclear power agreements, and collaboration schemes cannot be lost sight of. With a population exceeding 600 million people and a growing geopolitical role in Asia, economic considerations may perhaps be relegated to lesser importance than purely political factors. In the interests of the country, therefore, self-sufficiency in energy even in cases where not fully justified on purely economic grounds, is an important goal, which government and society must strive toward in all their actions. The costs of achieving this goal can, however, be minimized, by establishing mutuality of interests between those countries that are currently supplying energy in various forms to India until such time as indigenous sources can be developed at

CONCLUSION

reasonable competitive costs. The challenge posed by the energy sector is both exciting and formidable, and researchers and policy makers are likely to be engaged in meeting it for years to come.

ABOUT THE AUTHOR

Dr. R. K. PACHAURI is senior professor at the Administrative Staff College of India at Hyderabad. He has wide industrial, research, and teaching experience, both in India and North America. He was earlier Assistant Professor and currently teaches summer school at North Carolina State University at Raleigh.

Professor Pachauri has the degrees of M.S. in industrial engineering and Ph.D. in economics, as well as industrial engineering, from the North Carolina State University. The author has written <u>The Dynamics of Electrical Energy Supply and Demand: An Economic Analysis</u>, published in 1975 by Praeger, and has another book of edited papers on India's energy policy that will be published toward the end of 1977. He also has to his credit a number of published papers, as well as papers presented at various professional seminars and meetings.

RELATED TITLES
Published by
Praeger Special Studies

CHINA'S ENERGY: Achievements, Problems, Prospects
    Vaclav Smil

OIL IN THE ECONOMIC DEVELOPMENT OF VENEZUELA
    Jorge Salazar-Carrillo

RESPONSES TO POPULATION GROWTH IN INDIA: Changes in Social, Political, and Economic Behavior
    edited by Marcus F. Franda

ARAB OIL: Impact on Arab Countries and Global Implications
    edited by Naiem A. Sherbiny, Mark A. Tessler

Soc
HD
9502
I42
P32
1977